Teaching,Learning

and

Mathematics

Contents

Acknowledgements

(A)   My Classroom, by Anne Watson
Previously unpublished.

(B)   Reflections, from "In our Classrooms" ATM (1993) Contributors: Laurinda Brown, Margaret Jones, Jools Outridge, Mike Ollerton, Geri Rowe, Michelle Selinger, Anne Turnbull, Charlie Wall, Anne Watson, Aled Williams, James Doyle and Erika Pye.

(C)   The Impact of Beliefs on the Teaching of Mathematics, by Paul Ernest,
from "Mathematics Teaching: The State of the Art" (1989) Published by Falmer Press.

(D)   Towards a Language of Struggle, and Comments on Language in the Classroom,
from "Language and Mathematics", ATM (1980) Contributors: Bill Brookes, John Dichmont, Eric Love, Judy Morgan, Dick Tahta and Jim Thorpe.

MT (Mathematics Teaching) and MM (Micromath) are both journals of
The Association of Teachers of Mathematics.

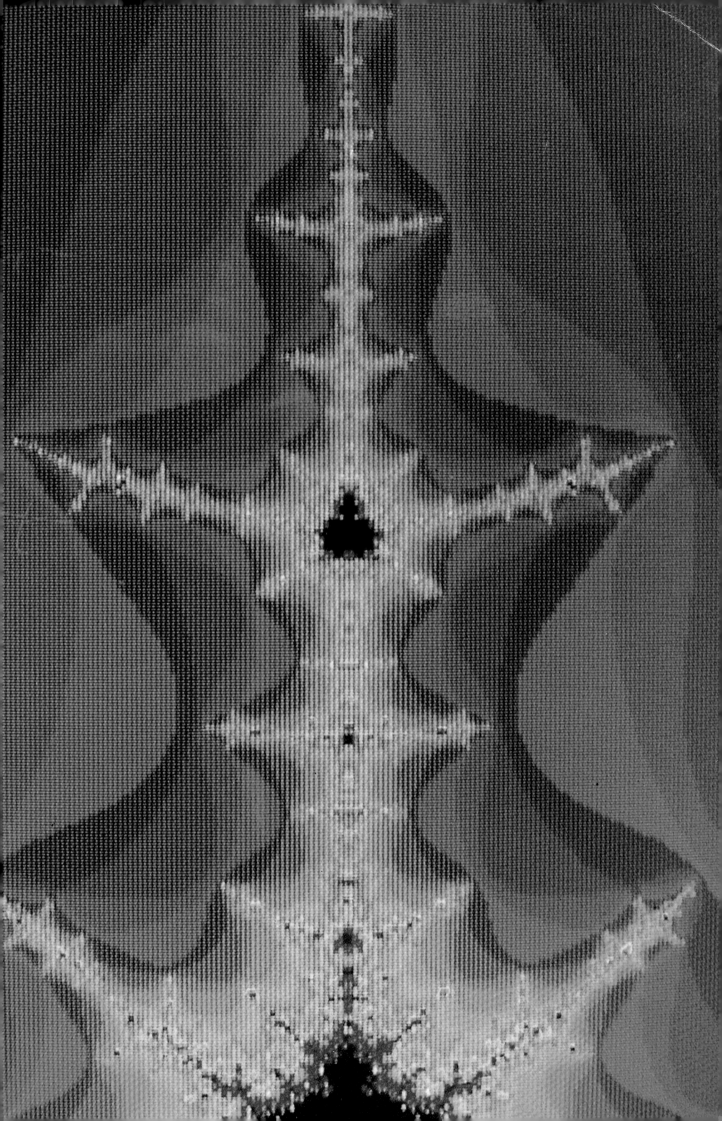

Taken from the ATM postcard set, 'Images of Mandlebrot', by Lyndon Baker.

REFLECTIONS ON TEACHING AND LEARNING MATHEMATICS

INTRODUCTION

In putting together this collection of articles we are aware of the potential audience, of our own involvement with ATM and of the essentially timeless nature of most of the writing.

At an ATM General Council Weekend a group of us, involved with initial teacher education, decided to compile a PGCE Reader from articles in MT and Micromath. We spent an hour or so, diving into past editions of the journals with: "We must have this one!" or "I'd forgotten all about this, but...". As you might expect this was an entirely non-critical phase and naturally by the end even the few rules we had were broken. Other ATM writings had appeared and even special pleadings for non-ATM material. There were even volunteers to write additional features for the PGCE reader. The next stage was to trial the material with postgraduate teachers. To be honest, many of us were in the position of having to produce such a reader anyway!

Secondary Initial Teacher Education is in a state of change. As we write, there is a shift of emphasis in how such courses are organised. With an increasing role for school-based training and a shift of resources to the schools, many universities and colleges are facing a period of uncertainty and mathematics educators are part of this process. However in reading the articles in this collection, we soon became aware that little of this current discontinuity seemed to impinge on their relevance. The writers were concerned with mathematics, the teaching of mathematics and the learning of mathematics in a way that made contemporary organisational matters irrelevant to our collection. Perhaps this was implicit in writing for ATM or perhaps our selection was influenced by our involvement with the Association.

Having collected the articles, it soon became clear that others might appreciate them, as well as teachers on postgraduate courses. many teachers in schools now play a larger role in the education of secondary mathematics teachers and would welcome such articles. Although many of these teachers will be members of ATM, others may be unaware of the Association's journals and the valuable part they have in learning to teach mathematics. Discussion on articles such as these can provide a means of encouraging reflection on one's practice and talking about it with new colleagues. They can supply a route into the shared reflection on practice, which is invaluable to both experienced practitioner and novice teacher.

We recognise that what follows is our selection and we are bound to have missed out certain articles which you may consider essential. We can only ask you to let us know of the omissions and assure you it is entirely an oversight on the part of the editors and not some machiavellian plot to undermine the teaching of mathematics. After all there may well be a second volume of articles and the sooner we start collecting the better.

ALAN BLOOMFIELD & TONY HARRIES

Chain, taken from the ATM postcard set, 'Spirals', by Lyndon Baker and Ian Harris.

SECTION 1: REFLECTIONS ON TEACHING, LEARNING AND MATHEMATICS

This section begins with "My Classroom" by Anne Watson. It can be a valuable stimulus to talking about the mathematics classrooms in your school and to considering why they are as they are. The six 'Reflections' on teaching are from 'In Our Classrooms', an ATM publication. Many new entrants to secondary teaching may find the concept, in the second reflection, of the "mathematical space in which I am trying to get the pupils to explore" a useful way of thinking about their planning of lessons. The next two passages provide clear messages about the joy of teaching mathematics. Jean Melrose's marvellous dream that "Mathematics teaching at all levels should contain opportunities for goggle-eyed wonder" contains the ambiguity that such wonder is available to the teacher as well as the learner. In his argument for the interplay between content and process, David Wells eloquently puts the case for "mysterious objects, amazing ideas, wizard wheezes and stunning insights".

Paul Ernest's "The Impact of Beliefs on the Teaching of Mathematics" proposes that three factors influence teachers' autonomy in practice; beliefs, social context, and level of thought. A model for teaching based upon the fostering of a conjecturing atmosphere is proposed by Alan Wigley in 'Models for Teaching Mathematics'. Both of these articles could be used as stimuli for an exploration of the factors and philosophies underpinning the teaching of both experienced and novice practitioners. Dave Hewitt points out that there is a need to develop an emphasis on the real mathematical content of what children are doing rather than the labelling that accompanies it, by spending time involved in the individuality of their work.

John Mason in 'Only awareness is educable' suggests that "only when I notice myself, do I become awake and free". Caleb Gattegno was a founder member of ATM. Gattegno's influence on a generation of mathematics educators has been enormous. Mason uses a quote from Gattegno as the title of his article. He argues that "children can do a great deal more with themselves than the most adventurous educator ever dreamed of". In recognising the inner power of children Gattegno warns of the myopia of adults looking at them from the outside. Richard Skemp's influential article 'Relational and Instrumental Understanding' was published by ATM in Mathematics Teaching in 1976. In the interview with Anna Sfard included here he argues that a "coherent and communicable theory" makes it possible to produce more teachers of high quality. The challenge is to capture the intuition of the excellent teacher before it disappears. In learning how to teach, discussion of our intuitive decisions and those of others is an essential element.

The end point of this section has to be David Kent's 'Linda's Story'. In considering factors which determine our approach to teaching, we have to recognise that our enthusiasm and the importance we attach to mathematics may well have side effects, which are more important than the learning outcomes we consciously aim for and often miss. Our behaviour can change others in ways we can never predict.

MY CLASSROOM

Anne Watson

When I tell people I am a mathematics teacher a common reaction is for them to recoil slightly and look at me with blankness or suspicion. I have, over the years, sometimes found the courage to respond by asking: "What do you think I do all day?" Sadly, but not surprisingly, the answers are nothing like the truth, but it is hard to know where to start with the truth. Much of what I do begins with an affection for children and mathematics, and some worked-out beliefs about learning.

The students

I believe that youngsters are naturally curious and want to learn. They learn from each other all the time and we should harness this curiosity and social learning to help them succeed in school. They are worried about being wrong and nervous about asking for help if "being wrong" and "needing help" have, in the past, been causes of low self-esteem by leading to ridicule, labelling or punishment. They like to participate, they long to be members of whatever is going on, but the teacher should give them ways to take part physically, socially or intellectually in the work. They need a job to do and their work to be valued.

The subject

If you agree with the preceding description of young learners, then I hope you will see that the teacher's job is to find some sort of balance between the demands of the subject, the syllabus and these natural human needs of the students. This is not a universally-held view, but in my experience it enables the teacher to be more encouraging, the students to achieve more and behavioural problems to be reduced.

If, on the other hand, mathematics is viewed as a linear progression of techniques which must be performed correctly then only a select few will be successful or enjoy it. It will not fit with the other ways students learn. It may still appeal to those whose intellectual activity is already in that mould or who, for whatever reason, already realise and understand the power of such knowledge. It might also appeal to those who are out of phase with their peers. Indeed many who enjoy mathematics say: "it was a way I could escape", "it made me feel different and special", "I liked the neatness and correctness of it". I think one can still organise teaching to appeal to these senses but also find ways to show others that mathematics is interesting by:

exploiting its internal patterns and symmetries;

using chunks of mathematics which interlink or occur in accessible places to motivate students to explore;

using things one can see or hear or touch to generate interest;

teaching the skills of enquiry, generalisation, exploration alongside those of calculation, estimation, manipulation;

and so on.

My beliefs are backed up by research but I came to them through my classroom work. I watch children and listen to them. I adapt my work to their responses. I use my knowledge of them as people to guide my teaching.

My classroom is a safe place. It is comfortable and pleasant and what goes on in it is caring and not abusive. But it is also a challenging place where students expect to work hard and learn mathematics. They are expected to participate and achieve whatever is within their power to do.

The Classroom

The classroom is a long rectangle with the two long sides mainly glass. One side looks out onto a corridor and the other a sunny courtyard.

The desks are arranged in L-shapes This encourages discussion but discourages chat or playing with feet or equipment. Children often need to check out their ideas with their friends and peers, especially adolescents, but the purpose of the room is work and the layout must clearly state the purpose.

There are posters and displays all over the place. Some have been there for years and I cannot bear to take them down, others are current and frequently replaced.

Some of the posters I have up are timeless. Each panel of glass facing into the corridor has an Escher poster on it. I know that children look at them and discuss them as they hang around outside. On the door is a sign which says "Mathematics : centre of excellence" and another which describes mathematics as a world of patterns and relationships. It contains a typographical error which I have left there because I can find out how many people read it by how many mention the error. Many do. On the windows which face the courtyard I have hung posters which are bright or startling so that, when the sun shines through, the room feels lively and exciting. The recent craze for black-and-white designs has led me to use several black-and-white posters and students ask about them: what do the designs mean? how are they constructed? where do they come from? can they buy them?

Around the room I have hung some quotations about mathematics or mathematically related ideas. Every now and then students add to my collection and I sometimes see them reading the words of Einstein, or Galileo, or Russell instead of working on their own mathematics.

From time to time there is a display of students~ work I rarely ask them to produce work specially for display because I have found that tends to encourage a rather superficial attitude to the mathematics and too much attention to the presentation. I am more likely to stick up a whole class's work without warning, either selecting good bits from each person's work or hanging up the whole lot in such a way that pages can be flipped through and looked at. The idea is that everyone's work can be valued in some way. I might even put notices next to it indicating what I think is good about a particular piece, what mathematics it illustrates or how good an explanation it is. Since the emphasis in my lessons is more on "writing down what you do" than "writing up what you did" these displays are not neat or decorated but are far more exciting for their informality and really give students a useful buzz of recognition. They also give others a realistic standard to aim for. Offered a range of work students of secondary age can choose which pieces represent something they could aim at and achieve rather than being offered, perhaps, one perfect or near-perfect example which might daunt weaker students.

The environment is, I hope, a lively and encouraging mixture of the achievable and the mysterious, the topical and the historical, the obvious and the intriguing, the expected and the surprise. A bit like mathematics.

What happens to the student who enters this space?

TWO LESSONS

I am not going to tell you about all the ways I have used to start some mathematical activity with a class. There are many. I will tell you instead about two recent lessons. Both were with an all-abilities group of 14/15 year olds. Both led to two-week projects which could have gone on for longer.

In a way they were similar starts. Students sat around me in a kind of loose circle and we talked. There were no papers or pencils, no note-taking or chalk. Just words.

A lesson about money management

I talked to them about how the outside world often judges school mathematics by whether people can do money sums or not. They agreed that this is what their parents and grandparents often talk about. I asked them about the people who work on tills in supermarkets and what would happen if they had to do all the adding up themselves. We agreed that it would take too long, lead to error, be impossible in the noise and with the interruptions even if the cashiers were super adder-uppers.

"O.K. but we ought to be able to manage our own money without calculators. What sort of calculations do you need to do to look after your money?"

I was genuinely wanting to find out what aspects of money handling were within their knowledge. It did not seem to me to be very sensible to talk about mortgages or investments if their own concerns were about catalogues and post office savings. I was also wanting to find out if they already kept track of their money or were out of control in the sense of never checking their change, unaware of what their purchases would probably add up to and so on. In fact they told me that most of them did use estimation quite often when shopping and also knew something about the interest their savings earn if they did not know the details.

Some of them suggested, and others agreed, that it was useful to be able to add up short lists of money yourself, and multiply it either exactly or approximately. Estimation would be better than just guessing for some things. It would also be good to know how interest works so you could choose the best

Saving schemes.

I was delighted with all this information and their response in general. We had spent 45 minutes on this and not one calculation had been done. I went away and wrote down some of the financial affairs of an invented person who was their age with their interests. I used what they had told me about their own incomes and expenditures and interests. Next lesson I asked them to construct a financial plan for the invented person. Not only did they work very hard on this but I actually enjoyed the lesson myself ... the first time I had enjoyed teaching about money!

The outcomes varied. For some students it was a great achievement to produce a realistic statement of income and planned expenditure which would not land them in debt. Others explored the complexities of a bank account which earned interest every month when they were told an annual rate. I expected a range of end points between these two as I try to offer tasks which lead each student to develop their own mathematics from their current state of knowledge.

A lesson about generalisation

I had a bag next to me with several prepared strings of coloured cubes in it. drew them from the bag one by one asking similar questions about each. The sequences of colours in the strings grew more and more complicated The questions grew more difficult to answer.

The first string was all yellow. I asked "what colour would the next cube be?" "what colour would the 99th cube be?" and "where would the 23rd yellow cube? These are easy, obvious questions to start with so I allowed them to shout out answers or asked individuals, just to get them to talk and engage in the work. "How would you convince someone else that you were right?" This is not such an easy question but worth attempting as it holds the key to articulating a general description of the sequence.

The next string was alternately yellow and grey. I asked the same questions.

The next was two yellow and one grey, the next three yellow and one grey, the next two yellow and two grey, the next consisted of three colours and so on.

Try it yourself. It becomes challenging quite quickly. For the harder sequences I did not allow people to shout out but instead asked them to think in silence and, if they found the current example easy, to imagine a more complicated one to work on in their own heads. I think that students who shout out answers, or teachers who encourage quick one-word answers and take them from the first person with a hand up, actually discourage others from thinking and taking part.

In the lessons which followed I asked them to concentrate on the general descriptions of sequences, using algebra wherever possible but always trying to use words first, as if they were persuading a younger sister or brother. Again the range of responses was very wide. The weakest students were able, with help, to describe odd and even numbers as $2n-1$ and $2n$. The strongest were using sigma notations and integer functions, again with help.

This starter was suggested to me by a reception class teacher who uses the simplest version of it with her 5-year-olds. She was astounded to hear that I thought I could use it with A~level students with m colours and $a_1, a_2, a_3 .... a_m$ cubes of each colour in the string.

POSTSCRIPT

You may read this and think "I could never do all this..." or "I am not sure that her approach is valuable". To the first comment I would say that you cannot expect to develop your teaching style in a short space of time. It takes a while, and an underlying sense of purpose, to become a teacher whose actions, environment and philosophy are a coherent whole. To the second comment I would say that it depends what you value. Performance indicators are good and continue to improve, I could certainly convince you with figures, but I feel it is more important that the atmosphere, effort and interest in my classroom continue until the students finally leave. That is what I have tried to convey in this paper. It is my opinion that teachers have to decide what it is they really value and then do everything they can to help students achieve.

REFLECTIONS

First

What I am trying to create in my classroom is an environment where children are working at mathematics. I call this environment a 'space'. Sometimes, as with matrices and transformations, almost the whole of the mathematics syllabus is in the space and at other times, as with the nine-pin geoboard activity, just particular ideas e.g. area and drawing skills. I know when I have found the space, for as long as the motivation lasts, because the children are busy working - challenged, asking and answering their own questions and finding ways of making their ideas clearer to themselves by thinking them through. The 1089 lesson I use with classes early on in their school career because I have confidence in my way of operating with this idea in trying to establish my way of working. Always there is the process; always the problems, questions and ideas providing the thrust for the activities even when I am being focused about the content.

Second

As a veteran campaigner of 2 years teaching, can I identify any ingredients in my classroom practice which are important? I like to think I promote exploration and the autonomy of the pupil. Here is my story.

At the preparation stage I usually try to address the questions: How can this area of mathematics be investigated? or How can this area of mathematics be consolidated? I then do some maths myself to try out the various ideas I come up with. This helps me get a feel for the mathematical space I am trying to get the pupils to explore. I then have to find some starters which will motivate my pupils to explore this space. I usually find my first attempts are rather structured. This is probably because of my desire that all pupils must explore all areas. A lot depends on my confidence level. This does seem to be higher when I do the investigating myself. It helps me to see what avenues a pupil might pursue as well as giving me an idea about how much fun the investigating might be.

Investigating is empowering for teachers too! The doubts still remain of course: what I regard as possible avenues might not correspond with what the pupils regard as possible avenues, likewise my sense of what is fun.

Next come questions such as: How shall I introduce the starter? How shall I keep the investigation going? How or when should the project be brought to a conclusion?

A story is a nice way to start sometimes though at other times I have found the pupils prefer to be given the starter in a rather pure mathematical form. I like pupils to discuss in pairs or groups what mathematical issues they feel the starter raises and what tasks they can invent for themselves. We perhaps collate these ideas and discuss them as a class so that each pupil can see the range of options which they could explore. I find this method supports pupils in deciding what mathematics they are going to investigate. I might also ask what resources or knowledge they think they need and give a quick refresher course where necessary.

Often different pupils are working on different tasks within the same mathematical space. My role for the next two or three or four weeks is to ask pupils how they are getting on: *Tell me what you know?* being a favourite of mine. I often ask pupils to share their ideas. This helps them to improve their mathematical communication as well as enlightening them to other approaches to a task or other mathematical tasks within the space.

I do like to retain some control and may take a vote on when to wind up a project and negotiate a deadline for reports or coursework to be handed in. Sometimes I just autocratically dictate my terms. At this stage I organise some consolidation activities which may be worksheets or games or a request that pupils make up consolidation questions for others to try.

Third

I try to intervene in a way which ensures that each

pupil has a critical moment in each lesson. By this I mean that each pupil should be at a different state of knowledge, understanding or confidence after contact with me.

I do this by looking for opportunities to enthuse about small breakthroughs or successes, even if they seem trivial to me. When I am leading discussions I try to winkle success out of pupils. I am the sounding board but I also nudge and make suggestions. The pupil should b the successful one in dialoguewith me. I do this whilst working with individuals or in front of the whole group. It is important for them to be successful in front of their peers: *Talk to me about your mathematics.*

Fourth

I demand enormous effort from my pupils. I ask them to take risks, ask questions, answer my questions without fear, explain to each other and challenge each other. In return I try to offer, either to a whole group or to individuals, starting positions from which they can explore and discover. I use a variety of stimuli: words, pictures, the written word, actions, demonstrations, expositions.

I do not expect the pupils to rediscover mathematics. They are frequently offered another way to look at things or a technique to help them. I have some idea of how one thing can lead to another and I try to judge when it is appropriate to show this. On the other hand I am not disappointed if they go off in another direction. It is not unusual for a class to have about fifteen different things going on in it. To make this bearable I urge independence and responsibility.

I look for opportunities to praise and share. I ask: *What are you doing? and Why?*

I changed my practice radically about 5 years ago. Changing was hard. The change process came about because of a feeling of inadequacy. I had not come to terms with some of the new developments that were around in mathematics education and felt I was being left out of something exciting. I was starting to read a lot and what I read began to make me feel as though I had never taught well in my life. Eventually these feelings passed and I was able to select the good points of my teaching and put them into a new context; a framework which contained some old and some new approaches to the mathematics curriculum most of which I felt comfortable with. I now try to teach children by finding out what they know and how they approach a problem by offering them as much autonomy as I can. I still experiment, but in the end I hope I encourage a reflective atmosphere by inviting pupils to articulate their conjectures, to explain what they believe they have understood and how they came to that understanding and to realise that being stuck or making a mistake can be valuable for their learning.

Fifth

Mathematics is interesting, challenging, frustrating, but should never be boring. In the classroom I try to provide an atmosphere in which the pupil wants to work on the mathematical ideas that are presented. In reflecting on my practice I am going to describe what I would like to happen.

There is in my head the ideal lesson, in which each student is engaged in a rewarding yet challenging experience. The reality quite often does not match with this dream for all sorts of reasons. The pupils are thinking of something else. The challenge is too challenging for them. The ideas are uninteresting to them. All too often I have to compromise and work with the experience the pupils are having rather than the one I wanted them to have, but this does not necessarily mean they are not working in a challenging situation, merely that the parameters within which I envisaged working have changed as I responded to the needs of the class.

If I have judged correctly the mathematics will be challenging rather than frustrating. The pupil who becomes frustrated also becomes disaffected and, for me, the interesting situations are the ones where I have to judge when the pupil is moving into a situation where they feel frustrated and at that point providing the right input/question which will move their thinking forward into a more positive frame of mind, in which they can begin to work productively with the mathematics.

I want the pupils to be discovering mathematical knowledge by using it in various situations, but recognise that there are occasions when I need to teach a specific piece of content, or need to set a 'page of sums' in order to practice a mathematical skill. But always my aim is to return to the situations in which the pupil is working with the mathematics, solving my or their own questions.

The recognition that the interesting questions are the ones you think you might be able to do, but aren't sure, is the first step towards the pupil becoming a mathematician. It is true that every child isn't going to become a mathematician, but they need to be able to experience the feeling that comes from having worked on a question, come up with a solution, or set of solutions that they are happy with, convinced that they have worked through the task in a logical way, able to justify or prove for themselves, and knowing that they are correct in the work they have done.

Sixth

Writing about my lessons it seems to me that much of my work is actually about emotion and relationships. In my classroom I try to create a situation where everyone really does feel challenged but none daunted. It has to be OK to make mistakes, including me, and no-one has all the answers. Sometimes we will meet problems we can solve today, others will have to waite and some may be insoluble. Our work will depend on the support we give each other. We work in a variety of ways: sometimes independently, sometimes in groups, sometimes investigating, sometimes from a text book. The task is always shared, we all know what we are aiming to do and we can negotiate what we do next.

Mathematically we are all learning. We can see our experience as a journey along a road and we know that some people have travelled further than others but, travelling further just takes effort and time and is accessible to everyone.

Strategies I use regularly are:
- taking time to explain the plan of what we are going to do today, tomorrow, next week and even next term/year
- setting pupils off on tasks which, although they are appear easy to start with, are actually very difficult whilst maintaining a policy of: *I wouldn't ask you to do it if it was easy and if we work together it will be OK*
- always giving answers with questions set from books
- responding to: *I don't know what to do* with *What are you trying to do?* and *How can I help?* or *What are you stuck on?*
- listening to what pupils say and more importantly what they don't say
- watching body langauge to know when to intervene with pupils and being prepared not to interrupt if my comments are not wanted e.g. when someone is engrossed in a task.

LEGISLATING FOR RAINBOWS?

Jean Melrose

It was good to read in the National Curriculum proposals of August 1988 that:

Mathematics is not only taught because it is useful. It should also be a source of delight and wonder, offering pupils intellectual excitement, for example, in the discovery of relationships, the pursuit of rigour and the achievement of elegant solutions. Pupils should also appreciate the essential creativity of mathematics: it is a live subject which is continuously evolving as technology and the needs of society evolve. (para 2.2)

It seems that the authors are describing what should be the case, what ought to happen in mathematics learning. It's not just the fostering of *a positive attitude to mathematics as an interesting and attractive subject* but the astonishment, silence and excitement felt with a breakthrough in insight. An equivalent image would be the wonder of a rainbow breaking through a dark stormy sky. But how? How does a good and conscientious teacher enable such things to happen? Rainbows are unexpected, they are a surprise and a delight but they are only likely under certain weather conditions. How does the teacher provide for the unexpected and the unpredictable?

Rainbows are infrequent but not that infrequent. Many contributors to MT have described such experiences. Consider these 3 situations with unlikely children making mathematical leaps that took me completely by surprise.

Colin, an amiable member of a lower set in the fifth year who ended up with a CSE grade 5. Only half of Colin's maths set turned up that afternoon. I checked the legitimacy of the whereabouts of the rest of them, looked round at the assembled company of 12, scrapped what I had intended to do, remembered the difficulties they had with areas and volumes of enlarged shapes (enlargement by a scale factor of 2 caused the volume of a solid to increase by 2, possibly by 4 or if they noticed my expression, by 6 times!) and found some 5mm squared paper. I found myself explaining that a 10×10 square depicted a field with just enough grass in it for 100 cows and could they please draw a square field with just enough grass in it for 200 cows? Introspection overcame me, why was I regressing to regurgitating

ill-digested bits of Piaget under pressure? It was hardly the most riveting and relevant approach.

But it had taken off. Gary had copied his diagram, a 20 by 20 square on to the board. Tracey had pointed out that *4 of them go in there*. Paul, finding confidence from somewhere, had drawn a 20 by 10 rectangle on the board. *It's got to be a square*. No marks for trying hard with this group! Paul's measure of acceptance by the group was that he was contradicted in the same way as everybody else. *It'll be 15*. They drew it and used various strategies for square counting to get 225. Paul had another idea and was keen to draw it on the board. He started drawing another 20×10 rectangle but realised part way through that he had been there before. *14 then, 196. Can we do decimals then?* I stopped thinking about the verb in that sentence and reiterated the problem. *The field must be square and it must have enough grass for 200 cows.* The group multiplying decimals were into an approximation to the second decimal place when I noticed Colin. His normally happy countenance had a serious look of concentration on it. He went pink, then red in the face, stood up and said with some violence *You put the bloody flaps on it.* Everyone and everything stopped. We were all trying to fathom what he had said. What flaps? Had he gone mad? He was still standing there — his idea was perfectly clear to him. The determination was still there. I preferred the chalk. *Come and show us what you mean.* He drew: *That's brill* was the unanimous verdict. It was too!

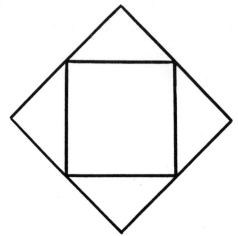

Jane, a shy, conscientious first year, who ended up with a CSE grade 3. Jane's class of 11-year olds were doing fractions. There was a spontaneous discussion amongst a small group about there being a lot of fractions between each pair of whole numbers. They repeated their argument to the whole class. There was general agreement that there were more fractions than whole numbers. I tightened the question a bit — *Were there more whole numbers or more fractions between 0 and 1?* and realised my convoluted language. *How many fractions are there between 0 and 1?* Opinion was divided. Millions was a favoured answer. If you drew a number line and started putting the fractions between 0 and 1 on it ½, ¼, ¾ ... then there would be millions of dots on the line. Infinity was suggested. I didn't want them to use the word as a label to stop them thinking about the situation: *What does that mean?*

It goes on forever. One of the 'millions' supporters said firmly that these fractions stopped at 1. Richard who was generally regarded as an able mathematician said *You can always find the average of two fractions and then find the average of one of the first fractions and your average, and then find the average of that average and the first average and ...*

– Will you always be able to do that?
– *Until you drop dead?*
– But someone else could carry on?
– *Yes.*

Most people were still concerned about whether, after putting millions of fractions in, the space between 0 and 1 on the number line would be filled up. They all wanted to talk at once, to argue with someone what they saw as undermining their firmly held convictions.

I got them to write down what they thought — just a paragraph. Richard calculated ¼, ¾, ½, ⅝, ⁹⁄₁₆, ¹⁹⁄₃₂... Many people were concerned about how blunt the pencil point was when you were putting the mark for the fractions on the number line.

Jane wrote: *You can make all the whole numbers into fractions and there are more fractions left.* Again there was the determination that I had seen on Colin's face the previous year. Again I was lost and unclear about what she meant.

She explained to the whole class:

You can make all the whole numbers into fractions.
You can make 2 into ½.
You can make 3 into ⅓.
You can make 4 into ¼.
You can make one million into one millionth.
You can make one billion into one billionth.
There will still be fractions like ¾, ⅔ left.
So there are more fractions between 0 and 1 than whole numbers.

There was the same quality of stunned silence as before. She had given a convincing argument to the millions contingent (my guess was that they had not seen the implications of Richard's work for their ideas) that their reasoning was incorrect and she had helped first ideas for form that the mapping x → ⅟ₓ will send all numbers greater than 1 into the interval (0, 1). She presented powerful new ideas that seemed to be growing in our imaginations. Negative numbers and the denumerability of the rationals could come later. It was a superb piece of mathematics.

Chris: I had no idea a 6 year old could create mathematics of such elegance

Chris was one of half a dozen 6-year-olds who were sorting plastic shapes — the animals set of resource materials seemed to have been mixed up with the transport set. The children were sorting by the criterion of what would go in the taxi, which was one of the shapes. The bicycle would, the car wouldn't, the aeroplane wouldn't, the cat would, the elephant wouldn't. The bird? Well that caused problems. How could you persuade a bird to be sufficiently amenable to go in a taxi? The discussion revealed that the children's attitudes could be sorted into categories 'hard authoritarian' and 'soft liberal'! The taxi wouldn't go into itself. That was daft, to think that it could.

The conversation ranged more widely as to what you could take into a taxi. Chris suddenly said: *I know what you can't take in a taxi. The world.* That quality of stunned silence again. Everyone mystified again. *I don't understand, can you explain, Chris, please,* I said trying to reflect our incomprehension.

If the world went into the taxi then that would mean that the taxi would have to go into itself! That sent me reeling! Sets mapped into subsets. Or were they classes? Something about Russell's paradox? I looked at the children. Chris had that determined look on his face that I'd seen before in Colin and Jane, the others in the group looked about as mystified as I did, a couple looked as if they had switched off.

Problems — with a 6-year-old both language and symbolisation for expressing such ideas is limited — answer — draw a picture!

The picture Chris drew was of a taxi and a boy, Chris himself, standing beside it. There was a spikey sun and a Cheshire cat grin on the boy's face. He was holding a globe (he'd been given one for Christmas) and the UK was marked and a tiny taxi drawn in just about the right place. The world was about to go into the taxi.

It is difficult to draw together common features of exceptional situations. These three seem to be characterised by:

○ having the space and time to drift a bit, not being dominated by the constraints of a scheme of work or syllabus. Any of these discussions could have been killed by declaring the subject matter not to be in a particular book or leading directly to a statement of attainment.

○ encouraging the pupil who has the insight to express what they are thinking and to communicate it to the whole group. They, in that determined state, are powerful and effective teachers. It's important not to get in the way of this process and it's salutary to reflect how often I must have limited or impeded good mathematics.

○ listening carefully to all members of the group and emphasising your common learning. *I don't understand. Can you explain?* needs to be common currency in these situations.

Colin and Jane are entirely enigmatic. They are not quite in the category described in the Cockcroft report:

However we have to recognise that there are some pupils who, even though they may glimpse the view from time to time as they become interested in particular activities, see in it no lasting attraction and remain indifferent or in some cases actively hostile to mathematics. (para. 7)

They were far from hostile to mathematics but they were not suddenly transformed after their insights into high attaining pupils in mathematics either. It may be that good ideas occur to all of us from time to time and in mathematics lessons a lot of them are not recognised.

The problem remains. How do we organise our implementation of the mathematics curriculum so that we and the children see the rainbows when they occur and appreciate their beauty? A recurrent dream of mine is that paragraph 243 of the Cockcroft report, that mathematical decalogue of the 1980s, actually starts

Mathematics teaching at all levels should include opportunities for goggle-eyed wonder... ■

University of Technology, Loughborough, Department of Education

WHY DO MATHEMATICS?

David Wells

What is attractive about mathematics? Wherein lies its magnetic quality? How does it hook your attention, and draw you in?

Mathematics has been variously described as mysterious, puzzling, suprising, weird and curious; elegant and beautiful; simple and complicated; displaying generality, unexpected connections, hidden depths, unity-in-variety; and last but by no means least, as extraordinarily powerful.

Some or all of these features may be found in the problems or situations that mathematicians start from, and in the objects, ideas and facts which they recognise or are familiar with, and in the processes and activities of doing mathematics.

(The term 'mathematician', by the way, is intended to include anyone doing any mathematics at all, from world-class professionals, to the small child arranging marbles in a pattern, to the newspaper reader whose attention is grabbed by a mathematical puzzle, whose mathematical content may be so cunningly concealed that the solver doesn't consciously appreciate that it is there!)

The process of solving polynomial equations up to the fourth degree, and the suspected impossibility of solving fifth degree equations in the same form, was once mysterious and puzzling, but is now clarified by Galois Theory which is elegant and beautiful, simple and powerful, and reveals unexpected connections.

On the other hand, the famous equation $e^{i\pi} + 1 = 0$ is itself regarded as beautiful by many, though, as I have recently discovered from distributing a questionnaire which asks for twenty-six theorems to be given marks out of ten for beauty, it is not difficult to find mathematicians who think very little of it!

The theorem that every prime number of the form $4n+1$ is the sum of two integral squares in just one way seems to me to be exquisitely beautiful. However, David Singmaster marks it down somewhat because it lacks any simple proof. An interesting interplay appears here between the attractions of a theorem and the attractions of its proofs.

Roger Penrose [1] records his delight on looking up the original proof of Ptolemy's theorem, and discovering *the beauty and simplicity of the argument, which just involved drawing one unexpected construction line*. Here it is the method which he finds beautiful.

The well-known 'billiards' problem invites many questions, which can all be elegantly solved by the method of reflecting the billiard table again and again in its own edges, reducing the path of the ball to a straight line and the original situation to that of the 'diagonals of a square grid' problem. This method of solution seems also to have a wide appeal, perhaps because it is so surprising. However, it is not necessarily the simplest [2]. Which do you prefer, a stunning, sensational solution, or a simple 'Why didn't I think of that?' solution?

Many mathematical objects are strikingly beautiful, to mathematicians of all kinds, and even to those who would claim (!) to be completely non-mathematical. A regular dodecahedron seems beautiful, in its regular simplicity. Five cubes inscribed in a dodecahedron will also delight many people, without demanding any explicit mathematical knowledge at all — only our brains' natural capacity for enjoying patterns.

Möbius bands and Klein bottles are intriguing, though their beauty lies more, perhaps, in the conceptions they embody than in their physical form. Even to mathematicians who are completely unimpressed by claims that the Golden Ratio is an essential component of the Great Pyramid and Botticelli's *Primavera*, the Golden Ratio may appear beautiful by virtue of its properties. For example, its continued fraction is the slowest of all continued fractions to converge to its limit. Numbers such as $\emptyset$ are attractive through properties which may not be immediately apparent, rather than being perceived more directly as visually beautiful.

The examples mentioned so far might be described as pure rather than applied. When mathematics is applied to solving problems in engineering or technology, or physics or chemistry, the criteria used and appreciated change somewhat. This is natural, given that criteria such as time or cost-effectiveness become important.

In particular, the interpretation of simplicity may change, and the significance of power. A powerful and simple approximate solution, which fails to illuminate the depths of the situation, may be far preferable to an elegant and illuminating general solution which cannot be put on a computer.

In physics and the other pure sciences, those that make most use of mathematics, the role of aesthetic criteria is evident from the statements of the scientists themselves. Thus Heisenberg wrote to Einstein [3]: *You may object that by speaking of simplicity and beauty I am introducing aesthetic criteria of truth, and I frankly admit that I am strongly attracted by the simplicity and beauty of the mathematical schemes that nature presents to us. You must have felt this too...*

Another example [4]: *I heard Linus Pauling walking along the benches of his chemical laboratory and addressing each project with, 'I like it, ... I like it ... I don't like it ... I like it.' Later he explained that man's aesthetic sense is his most profound intellectual capacity.*

Aesthetic criteria are important even in more mundane applications. I once discussed this point with a mathematician who had moved from solving difficult problems in industry into the local authority inspectorate. He felt that he was not so conscious of aesthetic feeling in the process of solution, but rather afterwards, when he could look back on the solution with satisfaction, including aesthetic satisfaction.

Do we want our pupils to appreciate these features of mathematics? If so, then they must meet them frequently, and in relation to every aspect of their mathematical activity.

Do the problems that they tackle grab their interest? Is Mary delving into the periods of decimals because she is intrigued by them, or because her teacher thinks she ought to be?

Is John spending ages over his problem because he refuses to let go of something that he is still making sense of?

Have the class been introduced to a wide variety of mathematical objects and images? Have they created many of them themselves? Do they realise that when they enjoy these objects and images, they may well be sharing their enjoyment with millions of people throughout the centuries?

Were they recently amazed by a mathematical object which they had never seen before?

Do they consciously strive to make their solutions as clear, and concise, and elegant as possible? Are they aware that solutions which seem elegant and beautiful are more likely to be effective, powerful, time-saving and pregnant with further possibilities – and of future use in other situations?

Do they share their work with others in the class? Are they aware that there is a wealth of intriguing, attractive mathematics which has been discovered and recorded by past mathematicians? Do they ever have the opportunity to share some of that

mathematics, to discover what it is about, to enjoy and admire it?

Traditional teaching paid no attention to affective/ aesthetic features of mathematics, and focused solely on getting standard results by standard means. This is hopeless. The few pupils who are attracted to mathematics taught from such a narrow perspective do so in spite of their teaching, not as a result of it.

However, an *investigative* approach which strongly emphasises process may, and often does, over-emphasise process, to the exclusion of every other aspect of mathematical activity. Thus, one authority [5], under the heading *Exploring with mathematics*, refers to investigational work, (the guide does not discuss problem solving,) like this: *The term 'mathematical investigation' has come to mean a type of work whose value lies more in the activity of solving the problem than in the solution itself.*

Such an approach is in danger of swinging from traditional domination by content to a limiting domination by process. This will be especially damaging if the process referred to is always the pupils' own processes, and never any of the fascinating and beautiful processes discovered by others.

Such an approach can easily eliminate from mathematics a host of fascinating facts, mysterious objects, amazing ideas, wizard wheezes and stunning insights – eliminating in fact a large part of what makes mathematics attractive.

There is a striking contrast here with science teaching. The most remarkable objects and phenomena of science tend to be well known – sometimes because we meet them in everyday life, sometimes because they get into the more-or-less popular press.

The science teacher has the benefit of this familiarity, and the good science teacher will exploit it at every opportunity.

Thus people are familiar with rainbows, snow and hail, homing pigeons, bees, elephants, treacle, roses, thorns, the Grand Canyon, Mount Everest, electrons, atoms, radio waves, magnetism, ... and often more or less familiar with black holes, quasars, psychotropic drugs, intelligent porpoises and so on.

Note here the vital point that pupils, or members of the public, do not have to be able to explain these phenomena in order to appreciate them. Rather, a significant level of affective appreciation comes first, while understanding, if at all, and only partial, may come later.

Mathematics teachers cannot rely on a natural familiarity with the most beautiful, weird and striking phenomena in mathematics, and have tended in the past to ignore them completely. The result is that pupils are deprived of the chance to wonder at the marvels of mathematics.

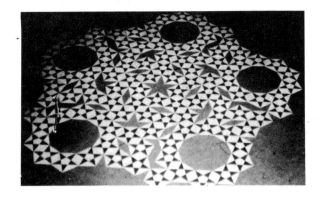

I recall my delight in a mathematics exhibition mounted by the sixth form at school and including many of the models in the then recently-published *Mathematical models* by Cundy and Rollett: it was the only event of its kind in my entire school life.

Nowadays, there are far more books of a popular nature which put many of the objects, facts and powerful ideas of mathematics within reach of pupils, including relatively young pupils. But it is not enough to rely on pupils' out-of-class reading. Opportunities need to be offered to pupils, preferably in an attractive and even dramatic form.

Such an attitude to mathematics, which treats it as a cornucopia of ideas and possibilities, of facts, objects and methods, is as far from emphasis on process as the essential feature of mathematics as it is from the traditional obsession with syllabus content.

Pupils should have the opportunity to appreciate the extraordinary activity of mathematics as a whole, from the fascination of investigating their own problems, to the delight of reaching their own satisfying conclusions, to the pleasure of appreciating its remarkable and beautiful objects, results and methods.

If they do not, they will not only develop a distorted view of mathematics, but they will be deprived of a large part of the enjoyment of mathematics, of the motivation for doing mathematics, and their ability to *do* mathematics will itself be damaged. That would be an ironic outcome indeed, bearing in mind the original intention of the promoters of open investivations and pupil autonomy − that pupils should become young mathematicians.

Content and process are not separate or separable aspects of mathematics [6]. Each involves the other, each feeds off the other, in a subtle and complex symbiosis. It is that delightful interplay that pupils need to appreciate, and as far as possible enter into themselves. ∎

Rain Publications

References
1 Roger Penrose, *The role of aesthetics in pure and applied mathematical research*, in Bulletin of the Institute of Mathematics and its Applications, Vol.10, numbers 7/8, 1974, p 268
2 For a simpler solution, see *Problem solving and investigations*, Rain Publications 1986/7, p 14
3 Quoted H Osborne, *Mathematical beauty and physical science*, British Journal of Aesthetics, Vo.24, 1984
4 H O Peitgen and P H Richter, *Frontiers of chaos*, Mapart 1985, p64
5 *Mathematics: GCSE: a guide for teachers*, SEC and Open University, Open University Press 1986, p 26
6 Process and content can also be interpreted as two extremes on a continuum, linked by general concepts which have both heuristic and factual aspects. See *General concepts and mathematics teaching*, in *Mathematics in school* 17-5, Nov. 1988, and *General concepts, and teaching*, in *Studies of meaning, language & change*, No.18, Rain Publications, March 1987
7 *The Scottish book: mathematics from the Scottish cafe*, edited by R Daniel Mauldin, Birkauser 1981, p 229

FROM THE INSIDE

Changing image

Mathematics subject assignment: *Share with us some mathematics in which you have engaged this term. Write an account of your findings and, at the end, write about the experience of carrying out the task – insights, frustrations, conversations, crises and satisfactions. How would you/have you introduce(d) the ideas to children? How does working on the mathematics for yourself inform your practice?*

Throughout my entire mathematical education this has been the first time that anyone has ever suggested to or encouraged me to look at some mathematics that interests me, and given me the responsibility of choosing it. In just about every other 'subject' I have been given choices in what I want to study or do eg humanities, art, CDT, science and health education projects. So, before I even started my assignment, the actual thought of it raised an important issue for me.

– Why, throughout my entire education, in which mathematics has always been my favourite subject, have I never looked at or chosen my own topic of study?

– Is it right that my mathematics has been restricted or limited to the choices my teachers or tutors have made?

– Is this difference in approach between subjects a contributory factor in many people's hate or misunderstanding of mathematics?

– Do people feel that, unless a teacher is present to teach or explain, then they can't do mathematics? ie maths cannot be 'done' outside the classroom?

Sara Stratton

On main school practice, The Ridings High School, Winterbourne

Students know when I am bored with a topic, or nervous, or uncertain. The unscrupulous can exploit these perceptions by being very effective at disrupting the proceedings. Somehow, they can find my weak spots quickly. If I am promoting a questioning, explorative approach, promoting mathematics as a disciplined form of enquiry and as a formalised expression of awarenesses that everyone who speaks and walks must necessarily have, then it seems to me that I must myself be in such a state when teaching. It is difficult to be genuinely surprised by an answer to a routine question that students have worked on year after year, but it is possible to be genuinely surprised and excited about the processes which students invoke, about what they have to say about their own thinking. I find myself getting excited in almost every session I run, and when I do not it is a disaster. But, my excitement is about being aware in the moment of recognising mathematical thinking; the use of imagery, the evocation of basic powers to make sense of things – and of possibilities to exploit what I notice and draw attention to mathematically significant moments.

John Mason

Centre for Mathematics Education, The Open University

THE IMPACT OF BELIEFS ON THE TEACHING OF MATHEMATICS

Paul Ernest

Official reports such as NCTM (1980) *Agenda for action,* and the Cockcroft Report (1982) recommend the adoption of a problem-solving approach to the teaching of mathematics. Such reforms depend to a large extent on institutional reform: changes in overall mathematics curriculum. They depend even more essentially on individual teachers changing their approaches to the teaching of mathematics. However, the required changes are unlike those of a skilled machine operative, who can be trained to upgrade to a more advanced lathe, for example. A shift to a problem-solving approach to teaching requires deeper changes. It depends fundamentally on the teacher's system of beliefs, and in particular on the teacher's conception of the nature of mathematics and mental models of teaching and learning mathematics. Teaching reforms cannot take place unless teachers' deeply held beliefs about mathematics and its teaching and learning change. Furthermore, these changes in beliefs are associated with increased reflection and autonomy on the part of the mathematics teacher. Thus the practice of teaching mathematics depends on a number of key elements, most notably:

> the teacher's mental contents or schemes,
>   particularly the system of beliefs concerning
>   mathematics and its teaching and learning;
> the social context of the teaching situation,
>   particularly the constraints and opportunities it
>   provides; and the teachers level of thought and
>   reflection.

These factors determine the autonomy of the mathematics teacher, and hence also the outcome of teaching innovations - like problem solving - which depend on teacher autonomy for their successful implementation.

The mathematics teacher's mental contents or schemes include knowledge of mathematics, beliefs concerning mathematics and its teaching and learning, and other factors. Knowledge is important, but alone it is not enough to account for the differences among mathematics teachers. Two teachers can have similar knowledge, but while one teaches mathematics with a problem-solving orientation, the other has a more didactic approach . For this reason the emphasis below is placed on beliefs. The key belief components of the mathematics teacher are the teacher's:

view or conception of the nature of mathematics,
model or view of the nature of mathematics
teaching,
model or view of the process of learning
mathematics.

The teacher's conception of the nature of mathematics is his or her belief system concerning the nature of mathematics as a whole. Such views form the basis of the philosophy of mathematics, although some teacher's views may not have been elaborated into fully articulated philosophies.

Teacher's conceptions of the nature of mathematics by no means have to be consciously held views; rather they may be implicitly held philosophies. The importance for teaching of such views of subject matter have been noted both across a range of subjects and for mathematics in particular (Thom, 1973). Three philosophies are distinguished here because of their observed occurrence in the teaching of mathematics (Thompson, 1984), as well as in the philosophy of mathematics and science.

First of all, there is the instrumentalist view that mathematics is an accumulation of facts, rules and skills to be used in the pursuance of some external end. Thus mathematics is a set of unrelated but utilitarian rules and facts. Secondly, there is the Platonist view of mathematics as a static but unified body of certain knowledge. Mathematics is discovered, not created. Thirdly, there is the problem solving view of mathematics as a dynamic, continually expanding field of human creation and invention, a cultural product. Mathematics is a process of inquiry and coming to know, not a finished product, for its results remain open to revision.

These three philosophies of mathematics, as psychological systems of belief, can be conjectured to form a hierarchy. Instrumentalism is at the lowest level, involving knowledge of mathematical facts, rules and methods as separate entities. At the next level is the Platonist view of mathematics, involving a global understanding of mathematics as a consistent, connected and objective structure. At the highest level the problem-solving view sees mathematics as a dynamically organised structure located in a social and cultural context.

The model of teaching mathematics is the teacher's conception of the type and range of teaching roles, actions and classroom activities associated with the teaching of mathematics. Many contributing constructs can be specified including unique versus multiple approaches to tasks, and individual versus cooperative teaching approaches. Three different models which can be specified through the teacher's role and intended outcome of instruction are:

Teacher's role	Intended outcome
1. Instructor	Skills mastery with correct performance
2. Explainer	Conceptual understanding with unified knowledge
3. Facilitator	Confident problem-posing and problem solving

The use of curricular materials in mathematics is also of central importance in a model of teaching. Three patterns are:
1. the strict following of a text or scheme;
2. modification of the textbook approach, enriched with additional problems and activities
3. teacher or school construction of the mathematics curriculum.

Closely associated with the above is the teacher's mental model of the learning of mathematics. This consists of teacher's view of the process of learning mathematics, what behaviours and mental activities are involved on the part of the learner, and what constitute appropriate and prototypical learning activities. Two of the key constructs for these models are: learning as active construction, as opposed to the passive reception of knowledge; the development of autonomy and child interests in mathematics, versus a view of the learner as submissive and compliant. If one uses these key constructs the following simplified models can be sketched, based on the child's:
1. compliant behaviour and mastery of skills model,
2. reception of knowledge model,
3. active construction of understanding model,
4. exploration and autonomous pursuit of own interests model.

Relationships between Beliefs and Their Impact on Practice

The relationships between teacher's views of the nature of mathematics and their models of its teaching and learning are illustrated in Figure 1. It shows how teacher's views of the nature of mathematics provide a basis for the teacher's mental models of the teaching and learning of mathematics, as indicated by the downward arrows. For example, the instrumental view of mathematics is likely to be associated with the instructor model of teaching, and with the strict following of a text or scheme. It is also likely to be associated with the child's complaint behaviour and mastery of skills model of learning. Similar links can be made between other views and models, for example:

mathematics as a Platonist unified body of knowledge - the teacher as explainer - learning as the reception of knowledge

Figure 1. Relationships between Beliefs, and Their Impact on Practice

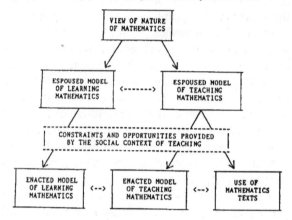

mathematics as problem-solving - the teacher as facilitator - learning as the active construction of understanding, possibly even as autonomous problem-posing and problem-solving.

These examples show the links between the teacher's mental models, represented by the horizontal arrows in Figure 1.

The teacher's mental or espoused models of teaching and learning mathematics, subject to the constraints and contingencies of the school context, are transformed into classroom practices. These are the enacted (as opposed to espoused) model of teaching mathematics, the use of mathematics texts or materials, and the enacted (as opposed to espoused) model of learning mathematics. The espoused-enacted distinction is necessary because case studies have shown that there can be a great disparity between a teacher's espoused and enacted models of teaching and learning mathematics (for example, Cooney, 1985). Two key causes for the mismatch between beliefs and practices are as follows.

First of all, there is the powerful influence of the social context. This results from the expectations of others including students, parents, peers (fellow teachers) and superiors. It also results from the institutionalised curriculum: the adopted text or curricular scheme, the system of assessment and the overall national system of schooling. These sources lead the teacher to internalise a powerful set of constraints affecting the the enactment of the models of teaching and learning mathematics. The socialisation effect of the context is so powerful that despite having different beliefs about mathematics and its teaching, teachers in the same school are often observed to adopt similar classroom practices.

Secondly, there is the teacher's level of consciousness of his or her own beliefs, and the extent to which the teacher reflects on his or her practice of teaching mathematics. Some of the key elements in the teacher's thinking - and its relationship to practice - are the following:

  awareness of having adopted specific views and assumptions as to the nature of mathematics and its teaching and learning;

  the ability to justify these views and assumptions;

  awareness of the existence of viable alternatives;

  context sensitivity in choosing and implementing situationally appropriate teaching and learning strategies in accordance with his or her own views and models;

  reflexivity - being concerned to reconcile and integrate classroom practices with eliefs, and to reconcile conflicting beliefs themselves.

These elements of teacher's thinking are likely to be associated with some of the beliefs outlined above, at least in part. For example, the adoption of the role of the facilitator in a problem-solving classroom requires reflection on the roles of the teacher and the learner, on the context suitability of the model, and probably also on the match between beliefs and practices. The instrumental view and associated models of teaching and learning, on the other hand, require little self-consciousness and reflection, or awareness of the existence of viable alternatives.

Mathematics teachers' beliefs have a powerful impact on the practice of teaching. During their transformation into practice two factors affect these beliefs: the constraints and opportunities of the social context of teaching, and the level of the teachers thought. Higher level thought enables the teacher to reflect on the gap between beliefs and practices, and to narrow it. The autonomy of the mathematics teacher depends on all three factors: beliefs, social context, and level of thought. Beliefs can determine, for example, whether a mathematics text is used uncritically or not, one of the key indicators of autonomy. The social context clearly constrains the teacher's freedom of choice and action, restricting the ambit of the teacher's automony. Higher level thought, such as self-evaluation with regard to putting beliefs into practice, is a key element of autonomy in teaching. Only by considering all three factors can we begin to do justice to the complex notion of the autonomous mathematics teacher.

References

COCKCROFT, W.H. (1982) *Mathematics Counts*, London, HMSO

COONEY, T.J. (1985) *Journal for Research in Mathematics Education*, 16, 5, pp. 324-36.

NATIONAL COUNCIL OF TEACHERS OF MATHEMATICS (1980) *Agenda for Action*, Reston, Va., NCTM

THOM, R. (1973) in HOWSON, A.G. (Ed.), *Developments in Mathematical Education*, Cambridge University Press, pp. 194-209.

THOMPSON, A.G. (1984) *Educational Studies in Mathematics*, 15, pp. 105-27.

MODELS FOR MATHEMATICS TEACHING

The job of the teacher is to make it easy for the students to learn. Or is it? **Alan Wigley** invites us to take a closer look at the curriculum we offer to learners of mathematics

The current scene

One potential advantage of a National Curriculum is that, with the content at least partly specified and ordered, we can move our energies from consideration of the *what* to consideration of the *how*. It is a challenge for teachers to work together on effective ways of approaching chosen topics. How this is done is likely to have a more lasting effect on pupils' learning and their attitudes to the subject than the particular content selected. Perhaps, since of its very nature a national curriculum cannot be idiosyncratic and must compromise, it will seem rather conventional to forward-looking teachers. This does not make me pessimistic. With some teachers at least, I sense an emerging re-orientation in which a chosen published scheme, rather than defining the course to be followed, is being used more selectively to meet national curriculum requirements. This opens the possibility of teachers taking greater control over what they offer their students, potentially to the great gain of the latter.

In a previous article (MT132) I suggested three major issues on which we in ATM ought to be working during this decade:
  (i)   resolving the content/process dichotomy;
  (ii) developing ways in which pupils can be helped to reflect on their learning;
  (iii) removing the unfortunate polarisation of the teacher's role into that of either instructor or facilitator.

These issues are germane to the whole educational debate and are not confined to mathematics. Hence the need to discuss them with fellow professionals and others with a public interest. More particularly, and despite changes over the years, I believe that we are far from achieving a consensus about approaches to teaching within the mathematics education community itself. Why, for example, do many authors of texts take on the impossible task of trying to create the whole

context for the learner on the written page when, as I believe, mathematics must necessarily be created 'in the air' of the classroom? I long to see more straightforward treatments of mathematical topics, enriched by descriptions of the historical and cultural context of the subject and with appropriate challenges for the reader. Lets have more dictionaries and reference books! Even the old-fashioned mixed bag of exercises is a good resource, particularly when pupils have to classify examples by type, sort out which they can solve and what methods are appropriate!

If we are to get our own house into better order, it is as important to tease out significant differences of interpretation as it is to emphasise similarities. In this article I shall first explore a model of teaching and learning which still seems to me to be too prevalent in mathematics classrooms. I shall consider some of the reasons why this model remains socially acceptable. I shall then describe an alternative model and indicate some of the ways in which it might be developed in the classroom.

The path-smoothing model

First, the main features of the model, the essential methodology of which is to smooth the path for the learner:

1 *The teacher or text states the kind of problem on which the class will be working.*

   The teacher or text attempts to classify the subject matter into a limited number of categories and to present them one at a time. There is an implicit assumption that, from the exposition, pupils will recognise and identify with the nature of the problem being posed.

2 *Pupils are led through a method for tackling the problems.*

   The key principle is to establish secure pathways for the pupils. Thus it is important to present ways of solving problems in a series of steps which is as short as possible, and often only one approach is considered seriously. Teachers question pupils, but usually in order to lead them in a particular direction and to check that they are following.

3   *Pupils work on exercises to practise the methods given aimed at involving learners more actively.*

  These are usually classified by the teacher or text writer and are graded for difficulty. Pupils repeat the taught processes until they can do so with the minimum of error.

4   *Revision*

  Longer term failure is dealt with by returning to the same or similar subject matter throughout the course.

Although this model emphasises *repetitive* rather than *insightful* activities, almost all teachers who use it as their basic approach will also consciously offer some insightful experiences. They will, for example, attempt plausible explanations, or encourage pupils to gather data about particular cases before offering a generalisation. However, there is usually a pressure of time felt by teachers, and consequently by their pupils, to move on to the 'work', which is perceived as doing exercises. The teacher may find the time to offer explanations but not to provoke the debate needed to clarify meanings. Inevitably, pupils' perceptions remain unexamined if they passively agree to the arguments in order that work can proceed. So attempts to justify and explain, although genuine in intent, can fail to convey understanding to the pupils.

It is important to note that the model is perpetuated in most textbook schemes. Individualised schemes almost inevitably follow the model, because they are dependent on the pupil being able to take small manageable steps, without constantly referring to others. So does any approach which basically uses a sequence of pre-structured questions and does not give pupils the space to explore their own responses to situations or to participate in making significant choices for themselves. We even have structured investigations, which attempt to reduce exploratory work to a series of algorithms or pattern-spotting exercises!

Teachers who hold to this model of teaching and learning certainly exercise professional care for their pupils and help them to achieve an important measure of success in public examinations. This care is shown in a variety of ways: by providing a structured framework with a clear work pattern, by marking the pupil's work on a regular basis and explaining where the pupil has gone wrong, by being available to sort out difficulties as they arise. As to public examinations, these tend to fall into a set pattern over the years and are therefore often amenable to a path-smoothing approach. The model is also one which parents and the public can recognise – a popular, if only partial, view of how learning takes place. These, perhaps, are the pre-eminent reasons as to why the model is so persistent in the face of an increasing body of knowledge and understanding about learning which tells a more complex story.

But limitations of the model may emerge in various ways. Sometimes learners flounder when presented with an unfamiliar problem, perhaps because they lack the strategies to explore the problem or the insight to recognise how it relates to problems which they have met before. From adults we hear comments such as 'I was no good at maths' or 'I could do the maths but never really understood what it was about'. On the one hand, we have a story of repeated failure and on the other, a lack of insight into mathematical relationships and their application to different contexts which has been only too prevalent in the adult population. Unfortunately being 'no good at maths' has itself become socially acceptable in some quarters, thus further perpetuating the model.

Some problems in teaching and learning mathematics

Before setting up an alternative model, I want briefly to consider some general issues.

There is a tendency in debate to polarise teaching and learning styles into one of two camps:

exploration	instruction
invented methods	given methods
creative	imitative
reasoned	rote
informal	formal
progressive	traditional
open	closed
process	content
talking (pupil)	talking (teacher)
listening (teacher)	listening (pupil)

The standard response is to declare oneself to favour a mixture of methods, neither entirely didactic nor entirely exploratory. But it is precisely here that the danger of a cosy consensus lies. The problem is not *whether* one should use a mix of methods (of that I have no doubt) but precisely *how* the blend should be achieved. Juxtaposing some open-ended tasks alongside a more prescriptive approach to much of the syllabus is not the kind of mix I have in mind! Neither is the path-smoothing model, even though it might sometimes blend the two approaches. When it comes to implementing an alternative model (what I shall call the challenging model), perhaps the chief demand on the teacher is to learn to live consciously with the creative tension which exists between exploration and instruction – in deciding, for example, whether to tell or not to tell about a particular matter.

A specific problem for teaching and learning

arises from the complex nature of mathematics itself. One aspect relates to the external 'meaning' of a mathematical idea, for example, what a 'half' of something is. It is important to be able to connect mathematics with aspects of the environment if we are to understand the subject and apply it to practical contexts. However, it is not as simple as that. Quite a lot of mathematics consists of invented systems with internally consistent sets of rules which cannot be abstracted from the world around us; the place value system for representing numbers is an example. Furthermore, to be competent in mathematics means partly to be able to operate fluently within the system, without constant reference to external meaning. For example, from an early age children build up experience of properties and relationships within the number system. Later, they learn how to manipulate algebraic statements, which are themselves generalisations of numerical relationships and thus a further step from 'reality'. Mathematical terms may have a multiplicity of meanings, with chains of interconnections. Thus a fraction can be interpreted as a part of a whole, as a ratio of two quantities, as one quantity divided by another; it has equivalent fraction, decimal and percentage forms.

Given this complexity it is small wonder that there has been an over-emphasis on rule-giving methods of teaching. How do you help students to develop their mathematics in meaningful ways? At what point is it appropriate to ignore other aspects in order to develop fluency within the system itself? And how do you do this without losing a sense of meaning and without placing an excessive burden on the memory? That is the challenge.

For many years there has been a feeling around that too many students are required to jump through hoops and are not able to think problems through for themselves. Particularly since the sixties, this has led to a search for better explanations or models for mathematical concepts which can be presented to pupils to help them understand. But this is not always effective. I will take just one example. If pupils fail to see the connection between the manipulations they make with the base 10 blocks and the marks they make on paper, then what useful purpose have the blocks served? Perhaps what pupils need here is less of a contrived explanation and more direct experience of the working system. This might be achieved, for example, with a computer program which merely cycles like a simple digital counter, so that students can infer the structure of the system by observing and discussing the patterns within the digits as they repeat. Some images, and not just the geometrical ones, are particularly powerful and pervasive – the cycling counter, the number line, the tabular array....Some of the work done in recent years, and written about in various issues of MT, suggests

that the role of the teacher might be to offer carefully selected experiences, in order to evoke images which can be developed and discussed. In this way the explaining, rule-creating or whatever, arises through sharing and discussing the experiences of learners; it is not externally imposed by the teacher.

Learners have to re-interpret what is offered and make sense of it for themselves. This is what it means to say that learning is an active process. It is not sufficient merely to work on problems and exercises. Learners reflect on experience, construct and test theories. Only then can they successfully integrate their knowledge and apply it in fresh situations. What we need then is a model for classroom practice which engages the learner by fostering a conjecturing atmosphere. (I assume that this is what attainment target one of the National Curriculum is trying to address.) Let us now attempt to outline such a model.

An alternative – the challenging model

The main features of the model are:

1 *The teacher presents a challenging context or problem and gives pupils time to work on it and make conjectures about methods or results. Often the teacher will have an aspect of the syllabus in mind, but this may not be declared to pupils at this stage.*

   An important word here is challenge. *The problem must be pitched at the right level, not too difficult, but more importantly, not too easy. The challenge may come from the complexity or the intriguing nature of the problem and the persistence needed to make progress with it, it may come from the variety of approaches which pupils bring to it, or from attempting to resolve the different perceptions which pupils have of a shared experience.*

   A second important word is time. *It is crucial to give sufficient time for pupils to get into the problem – to recognise that it poses a challenge and that there may be a variety of approaches to it – so that discussion begins.*

2 *Out of pupils' working is established a variety of ways which help to deal with the situation.*

   Here the role of the teacher is again crucial – initially, in drawing out pupils' ideas, so that they can be shared within groups or by the class as a whole. At some stage the teacher may wish to offer some ideas of her own.

3 *Strategies which evolve are applied to a variety of problems – testing special cases, looking at related problems or extending the range of applications, developing some fluency in processes.*

   Sometimes the syllabus requires the learning of more formal processes. The stimulus for this may be a harder mathematical problem and may require exposition by the teacher. However, the pupil will have the context of previous work to which more advanced techniques can be related.

4 *A variety of techniques is used to help pupils to review their work, and to identify more clearly what they have learned, how it connects together and how it relates to other knowledge.*

Longer term failure is dealt with by ensuring that any return to the same subject matter encourages a different point of view and does not just go over the same ground in the same way. The model places a strong emphasis on the learner gaining new insights, and the time required for reflection is considered to be fully justified.

Thus, at the heart of this model is the challenge to the learner. The teacher's function is not to remove all difficulties but to present the initial challenge and then to support the class in working on it. This may be a stumbling block if, as is often the case, pupils have different expectations of teachers and do not feel that they are being appropriately cared for. There seem to me to be three essential elements in developing the necessary supportive framework: to encourage collaborative work, to set more open tasks relating to global or key aspects of the syllabus, and to provide ways in which pupils can reflect on what they have done and relate aspects together. To achieve a necessary measure of teacher involvement, it is often preferable that the whole class should be working within the same topic. However, this does not preclude groups of pupils selecting their own topic of study from time to time, particularly once an effective way of working has been established.

For all but the youngest pupils, where paired work is perhaps more feasible, it is possible to arrange for small groups of say four students to work collaboratively together. Each group may work on their own strategies to solve a common problem, on different aspects of a common topic or, sometimes, on different topics. Within this context the teacher can work in depth with particular groups, bring groups together to share ideas, and inject new ideas when required. Individual work will still occur, both as a contribution to the group effort and to meet individual needs. However, the greatest challenge to the teacher is to encourage good discussion within the groups and develop a climate in which pupils can see each other as a first resort for help and support.

It helps if chosen topics are based on a major area of the syllabus, focus on central ideas, and extend over a lengthy period of time – perhaps even up to half a term. A substantial task can thus be presented, in which pupils help to: define the problem; develop ways of tackling it; generate examples to test a theory or practise a method; predict and make generalisations; and explore

further applications. Moving from exploratory work into new areas of knowledge requires specific strategies, such as sharing ideas about a carefully posed and challenging problem. Direct input by the teacher will sometimes be appropriate. In this way of working, a few key starting points are needed, sometimes supported with written sheets. A wide range of texts and reference books can be dipped into as required.

Another essential feature is reviewing, which is the term I use to describe various reflective activities, in which pupils step back from the immediate situation, and consider what they have done and how it relates to other aspects of their learning. Reflective activities can take the form of talking and writing about the processes pupils have gone through, making posters and reporting to the class, drawing up concept maps of a topic, and sharing attainment targets. Reviewing motivates and informs future learning activities and fosters general study skills.

Concluding remarks

I want to stress that the challenging model can be applied to the conventional syllabus and examinations in conventional (well, fairly conventional!) classrooms. It does not require extravagant resources, pupils of special aptitude or teachers of rare ability. It cannot be prescribed by the government, nor can it be proscribed by them! The benefits can be better learning and more positive attitudes towards the subject. Having described the key features of the model, it needs to be developed and illustrated with a variety of examples. I would be pleased if readers of MT responded to this article by giving examples to illustrate the model in practice in particular classrooms. There is the need to develop both the strategy of the challenging model and the tactics which might be employed with specific mathematical topics. Should we be seeking to establish this approach more widely in classrooms by the end of the century?

Alan Wigley, Adviser for mathematics, Wakefield LEA.

INVOLVEMENT

Dave Hewitt

In the research paper *The practice of reason* [1], Corran and Walkerdine point out that, during their observations of young children from three schools, the motivation children had was to 'get it done'.

> '... [the pupil's] satisfaction appears to arise from having done a quantity of work, no matter whether it was enjoyable, interesting or useful.

> This is how the children were motivated in relation to mathematics, and this was not confined only to those who found it difficult. Even the children who were successful at mathematics approached it with this attitude, ...'

Reading this, made me aware of some situations that happened in my classroom. I had recently moved schools and the children and I were not yet used to each other.

A third year class were involved in investigating a particular problem when a girl, Sarah, came up to me and almost threw her book on my table. I looked at her for some time, expecting her to say something. There was only silence and she was not even looking in my direction. (The conversation that followed I am writing from memory so it is not a precise account.)

DH: Yes? What do you want?

S: There's my book.

DH: Yes, well, what about it?

S: There it is.

DH: Yes I realise that. Did you have something that you wanted to say?

(Sarah opened her book impatiently, pointed at the page she had been working at, and looked away again.)

DH: Well?

S: There it is.

DH: There what is?

S: There's the work I've done.

DH: Oh, what was it you wanted to show me then?

S: Is it right?

DH: Is what right?

S: That.

DH: I don't know. Now what is it exactly you've been doing?

S: What you told us.

The conversation continued along these lines with Sarah getting increasingly frustrated with me. No matter what I tried, she never referred to what was contained within her book, the content of what she had been working on, the mathematics itself. It was simply 'that'!

The children in this class have a history of working through a rigid scheme which involved specific pages from certain books or cards at their own chosen rate. I had noticed that a frequent mode of communication between pupil-pupil, pupil-teacher and teacher-teacher, concerning the work that was taking place, was through such phrases as 'blue book 4', 'level 6.3', ... Only occasionally did conversation take place concerning the mathematics itself. So maybe I should not have been surprised by the fact that Sarah would not talk about her work since, for her, it had not been part of the discourse within the mathematics classroom. Had she been involved in the mathematics before coming to me, or had she been intent on 'getting it done'? The conversation we had seems to imply the latter.

How can I attempt to improve the involvement of my pupils in their mathematics? Weening them off a diet of books and worksheets, by introducing lessons that require a more investigational approach, is not sufficient as the encounter above demonstrates. One need is to develop an atmosphere in my room where conversations about the content of what they are doing is the norm rather than the labelling that accompanies it.

'That!', 'bookwork', 'card 25', 'page 8', 'the worksheet', ... are all examples of labels associated in some way to the work children may have been engaged in, but they are detached and remote from where the real activity lay. Often children have come up to me and proudly shown me their books saying 'Look, I've written 4 pages today!' Am I meant to be pleased? What does writing 4 pages mean? They might just have big writing.

Even a mathematical term such as 'probability' tells me next to nothing. What I want is some detail about what they were actually engaged in. This may not end up having anything to do with probability, it may be that one pupil spent most of their time grappling with adding fractions and another with discovering how to put a fraction into a calculator. Conversation about these will help them keep in touch with the mathematics they were engaged in. It means that when I listen to what they have to say, I will learn something about them and not just the neutral words and labels. In this way, although the whole class may be working under some title such as probability, the way in which each person approaches their task and the problems they encounter can be seen to be particular to them and that they have something unique to say.

When Sarah used to work through books and cards, she checked her answers from the card or answer book and saw that everyone was meant to end up with the same. It was only this common universal 'right' answer that seemed to have any importance attached to it. If I spend my time involved in the individuality of their work, then it may help the pupils attend to what they know and what they are struggling with rather than the labels or answers.

Such phrases as 'frogs' and 'squares on a chessboard' are rapidly becoming part of the mathematics teacher's dictionary and are just as distant from the mathematics that may be involved as 'blue book 4'. They are purely labels too. So, in a move towards greater investigational work, we still need to make an effort to keep in touch with the actual mathematics with which each individual child is involved.

Likewise, there is the possibility of maintaining the value judgements that are implied within the existence of an 'answer book'. Is the answer to the question 'How many squares on a chessboard?' important? It is not so much getting the answer that is important as knowing how you would be able to get it. Journeys are important. When someone engages in an activity, do they find a challenge that is appropriate and stretching? Do they need to make use of abilities that they possess in such a way that their involvement in this brings an awareness they did not have before? A new awareness they can practice and make further use of in order to gain still newer awareness. What does it matter whether an answer to the original question occurs within the process? A consequence of involvement is the succession of minor and major challenges that consume one's attention, leaving the original question somewhere in the background, to be recalled when necessary.   ■

Priory School, Weston-Super-Mare

Reference
1   G Corran & V Walkerdine, *The practice of reason*, Leverhume Trust Project, vol 1: 1981

ONLY AWARENESS IS EDUCABLE

John Mason

'Only awareness is educable.'
Caleb Gattegno

Gattegno's assertion is challenging enough when construed as applying to students. What does it mean when applied to teachers and educators?

Change

To think of changing others is presumptuous; to work on changing myself may serve as a role model for others. To work on changing my self is essential if I wish to be able to help others to change.

To change how I respond in a given situation, I must notice, in the moment, the possibility of choosing to act differently.

To act differently, I must have available a different form of acting, perhaps picked up from watching someone else, or possibly from reading or discussing with colleagues.

Only my awareness is educable, in the sense that my power to notice can be developed and refined, and my noticing can be focused and directed. Only when I notice spontaneously, for myself, can I choose. Only when I notice my self, do I become awake and free.

Fragments

Contrary to the standard cliché, we do not, in fact, learn from our mistakes. How often have you said 'I won't do that again!' and then gone and done it?'

What does it mean to 'learn from experience'?

Experience is fragmentary. We re-call fragments, re-member them or in other words give them shape. Fragments can be woven into stories, whose complex structure assists in re-calling related fragments. Memory is made up of fragments linked to and structured by stories.

Disparate fragments, unconnected by a story, remain unrelated in memory. To help students make sense, I need to help them develop their own stories. My own stories structure my own fragments, but may or may not help others structure theirs.

The term 'fragment' can be made precise by using it to refer to the details of events which can be agreed by all participants and observers. Even though all experience is construed and constructed, some details can usually be agreed, while others are more variably interpreted. Fragments tend to be re-called atomically, in one piece. The details which follow or accompany re-membering are linked by association and story.

Understanding

Explanation is the weaving of stories to account for the concatenation of a sequence of possibly overlapping fragments.

Understanding consists of fragment-linking stories, which provide rapid access to relationships whenever one or more fragments are re-sonated. The quality of understanding is based on the coherence of the stories, the extent of resonatable fragments, and the complexity of the connections thereby activated.

Very often we move so quickly from fragment to story, from an account of what we recall, to accounting for what we recall, that we confuse the story for the experience. Confusion and Babel arise when we believe our own and other people's stories.

To act *as if* a story were valid, as if a conjecture were valid, is not at all the same as believing the story or the conjecture. But it is easy to slip from *as if* to belief.

To try out a new word, a new idea, a new gambit in the classroom, I have to act at first *as if* it is useful and effective, in order to get the feel. I can only become convinced in myself if I pretend, yet simultaneously observe my self, remembering that it is only *as if*. All too often mathematics students become so identified with the 'doing', whether it be using apparatus, machines, or just doing exercises, that they lose contact with the *as if* (the only thing that mattered was the doing), and hence with the real purpose of the doing. All too often teachers become identified with one way of coping, lose contact with the *as if*, and are unable to grow or develop.

Mathematics educators, experienced travellers in the mental world of mathematics, can help pupils by bringing to mind the *as if*, by evoking the pupil's undoubted inherent mathematical powers where necessary, by re-minding pupils what they could be attending to − the general in the particular, and the particular in the general.

A discipline of mathematics education

Disciplines can be characterised by the ways in which they set about expressing and justifying generality, the types of invariance that they attend to, and the domain in which they ask questions and propose solutions.

What is the domain of mathematics education? What are the characteristic ways of expressing and justifying generality? What are the invariants in mathematics education?

Mathematics education is not just a particular case of general education. Mathematics is characterised by the fact that the objects studied are mental, not physical, and by the manner in which assertions are justified. It is people who do mathematics, alone and in groups. They need each other, and they need to be alone with themselves. Mathematics education is characterised by the application of psychological and social ways of noticing to inform the action of awakening and evoking of pupils' mathematical awarenesses, and to release their inherent powers to manipulate and communicate mental mathematical objects. Only awareness is educable.

Assertions that people in such and such a context will act in such and such a way, (a form of social laws analogous to laws of physics), may be valid for a time, but their validity is not robust, for awareness develops, changes and produces new behaviour. Weaker forms of such assertions, in terms of propensities, or possibilities, end up as descriptions of particular events.

'Mere' descriptions are looked down upon as too particular, as ungeneralisable. Yet if a description is sufficiently graphic and vivid, then it can resonate with my experience, summoning up both vivid particular incidents or ideas, and strong but generalised images or awarenesses. Only awareness is educable. If your description 'speaks' to me, it is far more general than some assertion to which I am unable to relate because I cannot see the particular in the general.

Seeing the general in the particular, and the particular in the general is fundamental to human functioning. It is the means whereby we simplify and structure our experience so as to make sense of it. It is the mechanism of resonance, the educating of awareness.

Why then are these notes not filled with vivid descriptions and anecdotes of classroom and in-service events? Because my conjectures (masquerading as assertions), although general, are also particular, and are intended to strike a chord, to resonate fragments in your experience, to stimulate reflection.

Reflection and separation

To educate awareness, to help teachers and students develop themselves and their selves, requires
○ support for positive (non-judgmental) reflection;
○ support for noticing moments when they could have acted differently, or wished they had acted differently;
○ support for preparing themselves to notice similar possibilities in the future.

Two birds, fast-yoked companions
Both clasp the self-same tree;
One eats the sweet fruit,
The other looks on without eating.
Rig Veda

Effective reflection is not just dredging up fragments from the recent past, and it is certainly not judging past behaviour. It is the creation of an inner separation. It is the awakening of the other bird, in order to break out of identification with mathematical computations or details of teaching, freeing part of the self to notice and monitor, to suggest alternative action – in other words to change. Only awareness is educable. ∎

Mathematics Faculty, Open University

Jackie Cook, Knowle Infants School

CHILDREN AND MATHEMATICS

CALEB GATTEGNO

Much of what follows will require that readers act upon their sights to manage to change their outlook. No reading of what follows can make sense otherwise.

Both children and mathematics exist. Whereas not many learned people can say, "I know what mathematics is because I know so much mathematics," and most ordinary people will concede that they know so little about mathematics that they will not advance any opinion about it, most of us do not see that this also applies to what the word 'children' covers. The only redeeming difference is to be found in the fact that we have all been children and can vouch that some established truth about them was also true of us when we were children. In fact, the difference is so important that I shall be using it as I develop my argument to establish the significance of many facts.

The study of children is thought to be the job of people as specialised as are those who study mathematics. Those who study children are called child psychologists and those who study mathematics are called mathematicians. Though mathematicians in their activity produce mathematics, child psychologists only try to understand children and explain their behaviours. They can only reach conclusions which are as good as the instruments they use, and we are permitted to disagree with their conclusions if, for example in our own case, we find a counter-example to challenge these conclusions.

The main difficulty in the study of children comes from the fact that the instruments used for that study leave out the reality of the inner life and its demands in the here and now upon the person we want to know. We have looked at children from outside and found only what we could see, not what is there to see. We have developed methods of study which have yielded what is now currently taught to teachers who generally believe it is the truth about children. If it is not, would they want to change their views? I think so.

To the readers of this article, I must say more in order to enlist their cooperation. I shall do it in the form of questions they could have asked themselves had they been alerted to the possibility of the inefficiency of the view from outside that is predominant today.

Why is it that each of us had to learn again by ourselves what is in use in our environment like sitting, standing, walking, running, etc.? Why is it that it takes time to learn all these things? That we are clumsy in the beginning but soon manage to master the skills on our own? That we know how to lead ourselves through exercises we give ourselves so that the teacher in us remains in close contact with the student in us?

Why is it that we play, and play seriously, all the games we enter into spontaneously?

Why is it that we sleep for so many hours every day in the beginning of our life ex-utero, although we do not seem to be tired and needing rest? Why is it that we do not notice for months that people speak in our environment and that we take so long to enter the activity of giving the language of the environment the capacity to stimulate us? Why is it that children crawl when no one else in their milieu does it?

How do children discover that their grip on things is to be educated and improved and practised? Which exercises do they give themselves to be able to grasp objects so that they can hold them? Which are the attributes of words which make children who are learning to speak, know which are nouns and which are verbs, etc.? Are there special sensitivities involved in these fine intellectual distinctions? Which are they? How do they work?

What is implied in the proper use of pronouns? Do children recognise them early and integrate them in their own speech with ease and total comprehension?

Is the correct change in words to indicate awareness of plurality and singularity a straightforward awareness open to children of that age (from 9 months to 3 years)?

Why is it that the obvious 'stress and ignore' process – needed by every child to enter into contact with only a small fraction of the fields he relates to and uses so systematically many times every day – is not recognised by outsiders as the working of the mind we call abstraction? In other words, since we always need abstraction to stress and ignore at the same time, and it is available to all of us, why is it that abstraction is denied to children who clearly use it masterfully?

There are of course very many questions which will alert us to what we did as young children and to what the children we meet do with themselves to master the numerous skills required in life to cope triumphantly with challenges of all kinds. I leave to each reader, now alerted that something meaningful has been ignored by those who inform us about children's performance, to add further questions; together we shall have a very impressive list indeed.

Since I have a very extensive list, I know that children can do a great deal more with themselves than the most adventurous educator ever dreamed of.

Children spontaneously stay with problems. And they stay for as long as is required. They consider abstraction (the simultaneous use of stressing and ignoring) naturally as their birthright. They give proof that they know many concepts but, more than that, that they know how to generate them in their awareness and how to recognise them as representable by a word and as represented by an open class of elements to which new items can be added. Their mastery of the language of their environment in their tender years, whatever that language, tells us clearly that they can perceive mental structures as present in the mind, how these structures link to each other and how they affect each other.

Moreover, they live close to their powers of transformation and to their mental dynamics.

It follows that whoever is aware of the actual demands of human living will find that each of us must be equipped to meet them, to be on top of them, to use them instrumentally, and that a great deal of it we do early in childhood. The most dramatic is our self-education in the fields of perception and action, and particularly in our mastery of language attained generally before we are 3 or 4 years old. If we know how to grant to children what they have obviously been able to achieve, we should be in a very different relation to them and we would stop making the abominable mistakes we unconsciously keep on making for want of the sight of truth.

Children spontaneously transform. They spontaneously know that everything is in flux and that a correct description of their world would be in dynamic terms. They know intimately that nothing is ever seen again in the same light, from the same distance and the same angle, and that objects are classes of impressions which are defined with respect to each other and by their overlapping parts. They know spontaneously that conservation is not needed to make sense of a world in flux and systematically ignore it until there is a place for it in their perception of the world; for example, when quantity becomes an attribute of some qualities perceived within a field.

The example of conservation – made into a shibboleth in some circles of psychologists and epistemologists – is important here, for children tell us they do not need it to live intensely in their world of experience. But myopic adults consider it an achievement in intellectual growth, a quality of the mind acquired by proper educational means. The real structure behind equality and identity is the much more flexible one called, officially, equivalence. Equivalence is the proper way of looking at the world because things are made to belong to each other (in a class of equivalence) when we stress one or more attributes and ignore others. It is the stressing that unifies, but since the ignoring is also present, it serves to reopen a question closed by the stressing. As children, we live it all the time and we are confused when the so-called knowledgeable do not see what is obvious to us. Thus we are being conditioned to see things with only a part of our wits and assert half-truths as if they were truths. Then we no longer know that we can trust our perception of the world, the inner and the outer.

Man's mind is the generator of mathematics.
Mathematics is a mental activity.
Mathematical structures are mental structures.

To become aware of these structures and of their transformation into mathematical structures may be what those among us who become mathematicians do spontaneously but do not quite know what it is in them that gives them that label, or how they can reach it in themselves to tell others about it. The phenomenon of the recasting of the "whole" of mathematics (around 1940), by the French group of mathematicians known as Nicholas Bourbaki, tells us that if we become aware of the 'mother' structures present in the special complex structures that make the entities which we encounter in mathematics, we act at the same time as mathematicians and psychologists of mathematics. What Bourbaki did was to produce the sequences of theorems which were the core of mathematics as they appear under the lighting of structures defined in mathematical terms. But they could only reach these by introspection, a psychological device. Being mathematicians, they did not concern themselves with this finesse. But we can and perhaps must.

(A brief list of examples given here included: equivalence replaces equality in all arithmetical relations; equivalence classes associated with arithmetical entities; algebra always present in any relation; spatial relations blend dynamic perception and virtual action; computational competence follows from correct perception of mental powers and use of electronic technology.)

Every epoch is the starting point of something which is important only if it objectifies itself in a lasting manner. Our epoch may attempt to give our mental powers their place in the world. If children are fulfilled in their play let us give our studies the quality which motivates everyone in the same way as games do and maintain contact with awareness of our mental powers which translate themselves into our growth. Mathematics is a field that lends itself well to such a creative blend of personal involvement in matters that have a future.

AN INTERVIEW WITH RICHARD SKEMP

Anna Sfard

This is a shortened version of an interview which was recorded in December 1989 during a visit by Anna Sfard to Warwick University.

Professor Skemp, your book The Psychology of Learning Mathematics (1971), and your article Relational Understanding and Instrumental Understanding (1976) are widely considered to be among the corner-stones of the relatively young discipline called the psychology of mathematics education. What are your views about the current state of research and about future needs of the domain?

In the domain of the psychology of learning mathematics, we have come a long way in the last twenty years. But there is always further to go, since the more our understanding develops, the more new ideas begin to open up. In general, the discipline can, and probably should, develop largely through creative dialogues and discussions, such as – I was very happy to discover – my paper about relational and instrumental understanding led to. Very soon after its publication, several articles appeared which contained important and interesting reactions to my ideas. I tried to combine all these together in my later paper *Goals of Learning and Qualities of Understanding* (1979). The discussion took my own thinking certain steps forward. But the most important thing about this exchange was that instead of reinventing the wheel every time, these researchers went forward from something which they saw as a useful beginning. This I think is the best way of building knowledge. It happens only too often that everybody puts his bricks in a different part of the field, so we never have a wall or a house. In the present case, I put some bricks down, and then others put their bricks on top of mine. Then I rearranged their bricks just a little, to make a surface which could support my next bricks; and so on.

Do you think, therefore, that your own work can provide a paradigm, a general theoretical framework into which other people might usefully try to incorporate their own contributions?

I remember something I said in one of my presidential addresses to PME, and which seems to be still true: that we need some kind of macro-theory, within which people's individual contributions can fit and cohere. This is what we have usually in the natural sciences.

The general paradigm which I think could best serve us in the way you describe is not in my writings on mathematics education, but in my generalization of these in the book *Intelligence, Learning and Action* (1979). My theory of mathematics education is now embedded within my theory of education, and my theory of education is embedded within my theory of intelligence, which is biologically based. If you are asking why somebody is a mathematician, you may have to look into his personal history; but if you are saying 'Why is he capable of learning mathematics?', the answer is that he is a member of a species *homo sapiens* which has special mental abilities, and you must ask how does *homo sapiens* acquire these characteristics, which are unique among all species on this planet? My book *The Psychology of Learning Mathematics*, I am happy to say, has had quite a wide readership (in seven languages including Chinese and Japanese), because it addressed itself to questions which people were already asking. Even so, I think *Intelligence, Learning and Action* is my most important theoretical book. There are many questions which can't be answered within their own limited context. A wide viewpoint is needed, and this book offers a viewpoint of this kind. A simplified version of this, slanted towards its application to mathematics, is given in *Mathematics in the Primary School* (1989).

You seem to use mathematics as a context for much more general psychological investigation. What is it that makes mathematics such a good laboratory for inquiring into human thinking?

Mathematics is an ideal subject for studying the functioning of human intelligence, and it was my good luck that I chose this as my area of research when I moved from school teaching to university teaching. Having done so, I gradually came to realise that mathematics is the most pure and concentrated, and therefore also the most powerful, example of the *functioning* of human intelligence. My model of intelligence says how it functions, not how to measure it. I am not saying that how to measure it is altogether unimportant – you can measure your money in the bank, and how much of it you've got is important. But if you didn't know what money is for, what a great resource it is, and all the things you can do with it, it would be useless. You could starve in the midst of plenty if you didn't know that by taking your money to a shop you could get food. Therefore, what you can do with your intelligence is what my model is largely about. According to this model, one of the important features of intelligent learning is its adaptability. It is because of our high intelligence, and the survival value of adaptability, that man as a species has come to the top of the pile so far. (And he will have a better chance of continuing there if he can learn to re-direct some of the primitive instincts which he is still stuck with).

There is another important way in which mathematics exemplifies the functioning of the human intelligence. I see mathematics as a powerful and adaptable mental tool. Although we have many mental tools, there is no other which is adaptable to such a wide variety of uses. You can take almost any little bit of pure mathematics, and then you can make a list of its different applications that would go on, and on, – from physics to commerce, to navigation … Think, for instance, about the countless uses of the simple formula $a = b/c$.

It is also worth mentioning that mathematics provides a low noise situation for studying human thinking. If you want to form a new concept for the first time, you need to study it in a relatively low-noise situation. And this is why, when you are teaching children mathematics, the best initial context comes neither from every-day examples and the physical world, nor from projects and investigations. Rather, it comes from low-noise examples, such as multi-base materials, unifix cubes, matches or milk straws tied into bundles of ten. And then when the learners have formed the concept in a low-noise situation, they can use it in a high-noise situation. Let us go back now to the issue of cognitive research. Here, once you see the difference between intelligent learning and habit learning in the low-noise situation of learning mathematics, where the differences shows up very clearly, then you can see the distinction in other embodiments where previously you would have not been able to penetrate the noise.

What are, in your opinion, the most important ingredients of effective teaching? More specifically, what is the relative importance of good teachers, of good programmes and teaching materials, of pupil's own attitudes, of appropriate methods of assessment, and so on?

Well, I think the problem is that one needs several things to be right simultaneously, in a number of different universes of discourse. Because children in school are human beings, they are several things at once: social beings, intellectual beings, emotional beings. All these create manifold needs which have to be given attention and satisfied together. When they are not, it is probably the higher mental processes which suffer first.

Maybe a good starting point for understanding this would be my model of intelligence, since this relates cognitive, affective, and social aspects of learning. If we see learning mathematics as the construction of extensible, highly abstract, adaptable and powerful schemas – and I believe this is what mathematical knowledge is – my model distinguishes three modes of scheme construction.

The first mode may be observed at the beginning level, in primary school. Here, the child has opportunities of working with physical embodiments of mathematics and for testing predictions based on his mathematical models. (Let us not forget that construction includes both building and testing.) Many Americans talk about using manipulatives as if that was all that was needed. But as Marilyn Harrison well said, children do not learn from manipulatives. They learn from the activities which they do with manipulatives.

At the secondary school level I am not sure what might take the place of manipulatives. But a child is very much a social being, so one should provide an interactive learning situation where children can work together in cooperative groups and sharpen up each other's thinking by discussion and mutual help and support. This is the second mode of scheme construction. Primary schools have known how to work in this way for some time now. The new GCSE exams open the door for secondary school children to work in this way too. This is splendid.

Now for the third mode of schema construction. When one has coherent and well developed schemas, these can be the starting point for creativity to function. We can extrapolate existing schemas into new situations, and we need learning situations in which this can happen. There has to be a classroom environment favourable to this.

As the constructivists have rightly emphasised, my model says that only the learner can construct mathematics (or any other conceptual knowledge) in his own mind. It is very much an organic model, because it suggests that learning is like natural growth. A plant's development is a cooperation between the roots, which draw its nourishment from the soil, and the leaves which draw energy from the sun. The sun, the rain and the temperature provide the environment, the soil provides the nutrient. In a similar way, intelligent learning requires a combination of several factors. The school has to provide the environment – not a desert, not an Arctic waste. It has to provide a micro-climate within which intelligent learning can take place. It also has to provide nutrients in the form of specifically devised materials, and this is the soil from which mathematical learning can draw its nourishment.

What about the teachers? What is, in your opinion, the relative importance of teachers and materials?

I think each needs the other. No programme can be more influential than the teacher who implements it. But in the field of mathematics there are very few teachers who either have the time or the particular combination of expertise to develop really good teaching materials for mathematics. Curriculum design and development are very demanding and time-taking.

Excellent teachers can help children to learn maths from archaic teaching materials (such as can still be found in our schools). This is what happened to me in school. It is only now, all these years later, that I realise that the materials my teachers had were archaic. Because I've had some superb teachers, however, I was able to become a mathematician. In fact, their teaching was not based on the materials; rather, it was based on their own intuitive excellence as teachers. But if intuitions cannot be verbalized they cannot be shared, so they die with the teacher. Sometimes they may be caught, as I think by me they were. But I had to verbalize them into a coherent theory before they could be explicitly communicated.

Teachers of such high quality, who can be excellent in spite of inadequate materials remain few and far between and fortunate are those who have them. On the other hand a coherent and communicable theory makes it possible to produce more such teachers. And teaching materials which embody such a theory are a good vehicle for learning the theory.

Speaking about changes in teaching mathematics, suppose you were appointed a minister of education. What would your first step be? How would you, if at all, reform mathematics education?

Frankly, I wouldn't want this position. Political power is transitory. The power belongs to the position not the person and when a person no longer occupies the position he loses his authority. In contrast, authority of knowledge truly belongs to a person himself. It is not imposed by law, but

freely accepted, and may continue to influence people's thinking for centuries to come.

But suppose that under persuasion you did agree to become a minister – what would be your most important move, to improve the situation?

Education must be freed from politics, it must be child- and learner-oriented more than it is subject-oriented. Teachers must see it in the long-term perspective of the future needs of a child as an adult, which will include earning a living, but not just earning a living. I believe that by working in a way that puts the realisation of human potential above everything else, everything else that matters is much more likely to happen. If you put the small things first, then you never get beyond the small things. But if you put the most important things first, if you really care about the realization of the potential of human intelligence, then all the other needs and goals have a much better chance of being achieved. It's a long view, and it is very important that they are not under pressure to assume short-term attitudes. Such pressure may be created by external examinations and by the evaluation of teachers' own work through the results their pupils achieve at these exams.

School education needs to be seen as only the beginning of a life long activity. This is even more important than before in the present age of rapid industrial and technological change, where intelligent learning and adaptability are even more important. Schools need to be allowed and encouraged to take the time to build good foundations, and to ensure that children's early enthusiasm and confidence are preserved so that they continue into adult life.

I said earlier that education needs to be freed from politics. I suppose that no profession is entirely free, but other professions such as the medical profession are more able to withstand political pressures than teachers. Can you see teaching hospitals letting the Minister for Health prescribe their curriculum? Or the engineers, or the legal profession? I see the independence of the education profession as no less important for democratic freedom than that of the legal profession. But I don't see how this can happen until teachers have a coherent viewpoint based on a unified theoretical foundation. Educational theory is at present a cluster of four separate disciplines: philosophy of education, history of education, sociology of education and psychology of education. I am hoping that one day my new model for intelligence, learning, and action will become the foundation of a unified theory of education. ∎

Hebrew University, Jerusalem

LINDA'S STORY

DAVID KENT
Belper High School

About the humanising of mathematics education

D.K.   *Which subject did you dislike most?*
Linda: *Maths*
Cathy: *Maths*
Yvonne: *Maths*

That is an extract from a report published a few years ago [1]. I would suspect that if you took a random sample of people aged 18 to 20 today, and asked them the same question, then the mode of the responses would be mathematics. I guess that many people would agree with my suspicion. I am also fairly sure that we do not know why this is so. Certainly, when I asked Linda why she disliked mathematics so much, she was unable to throw much light on the situation. Her analysis was rather nebular, taking the form of:

It was boring.
I could not understand it.
None of us could do it.
The teacher never explained anything.

The tragedy, from a mathematics educator's point of view, is that Linda did not dislike mathematics. What she disliked was that which was served up at her school. The analogy seems to be something like offering a person a piece of best steak, which has been burned to a cinder, and then assuming that he or she does not like steak.

There is of course nothing wrong with disliking mathematics, or anything else for that matter. I myself do not like smoking cigarettes, I do not like driving fast, really dislike cooked apples and can see nothing attractive in fishing. But so·what; we all have our likes and dislikes.

The big difference between my dislike for, say, driving fast and Linda's dislike for mathematics is that I have made my decision after participating in the act. Linda never really got involved in mathematics and could only make her decision out of ignorance.

Giving things up, whether you like or dislike them, is fair game. As a younger man I developed a passion and some flair for bridge. I taught the game and played competitively at a fairly high standard. My partner and I did, in one match, beat the county pairs champions. Then for what seemed to be no real reason I stopped playing, stopped reading the bridge articles and stopped thinking about the game. Very simply I chose to do other things with my time. I think I can fairly say that I had bridge, I knew what it was about, I liked the game and was a good player. But I dropped it, opted out, made a deliberate choice not to do it.

My opting out of bridge and Linda's opting out of mathematics are poles apart, at the end of the 'opting out' spectrum in fact. When a person drops mathematics in the way I dropped bridge it is fair enough, and the mathematics educator's job has been done. But one would be right in suspecting that few people give up mathematics in this way.

The millions like Linda who claim to dislike mathematics are mistaken. They do not dislike it; it simply passed them by, and in a way that was at great personal cost to them all.

Some years ago I was desperately looking for something to write about. Suddenly finding myself as a student again was strange and the thought of writing a 30,000 word essay was a little daunting. Little did I know that the young girl thumbing a lift into Exeter was to supply me with enough to write about then, and now—nearly five years later.

As the conversation in my car unfolded, I became aware of the tragedy of Linda's story. School had meant nothing to her; nothing positive, that is. "The teachers," she said, "made us rebel. Told us to leave. Said we were a bad influence."

I suppose Linda was in some ways a rebel. Certainly she would not allow herself to be pushed around. School had been a dead loss and work after it not much better. In one job she had refused to do something her boss told her to do. He gave her an ultimatum, so she left, immediately. She had been in that job for precisely four hours.

Life after school had been that of a drifter. She and her friends had more or less given up any idea of work, living in the coffee bars and off their wits. It was easy for attractive eighteen-year-olds to get by. They would go to a pub and before long a few young men bought them drinks, a meal and took them to the dances, pictures or whatever. So a succession of twentyfive-year-old boy friends pay for the entertainment, while the dole money pays for the essentials—which are kept to a minimum by living at home.

Linda told me that she was going into Exeter to meet her friends for coffee. I offered to pay the bill, provided they would be prepared to talk about education. Their philosophy was working again; here was another young bloke offering to pick up the bill, so what if he had different motives from the others?

It was a captivating morning, so personally significant that I can recall the finest detail from my imagery. Cathy and Yvonne were not in the least bit surprised that Linda had found yet another man. Not paying was all too easy. They were surprised that we spent all the time talking about educational matters. They were even more surprised when I asked them if they would allow me to teach them a little bit about mathematics over the next few weeks.

None of us ever saw Yvonne again, but Linda, Cathy and I met once or twice each week from February through to June, at the University, at my house, in the pubs, on the cliffs at Budleigh Salterton, at their houses—wherever we could. We received some strange looks when I spread the Cuisenaire rods out across a table in the *Coach and Horses*! But there seems to be no reason why you should not teach mathematics over a glass of beer.

Our lessons had no local structure, but there was a global one. By that I mean that I was never sure what I would be doing next, but I knew something about the overall picture I was trying to create. The content of our work varied from fractions, curve-sketching, logic, the teaching classic 'how many squares on a chessboard?', binary arithmetic, modular arithmetic and some rather advanced work on set theory. I also told a few stories from the history of mathematics. The movie men have found the stories of many of the great musicians and artists attractive box office material. Perhaps they should look at the potential of Galois, Abel, Boole, Vieta, Wittgenstein, Ramanjuan and others involved with mathematics. There are many romantic stories which we could all make use of.

My course of study finished and I had to leave Linda and Cathy. We wrote to each other for a few weeks and my family received a card from the girls for the first Christmas after. But, as is usually the case, the flow of correspondence stopped.

About four years later business took me back to Exeter again. Having half a day to spare I decided to ride into Exmouth (on my motorbike) and get in touch with Linda.

I telephoned her home from the phone boxes outside the post office. I was shocked when she replied.

"What are you doing here?" she asked.

"Long story, but how about a drink?"

"Sure, I'm just coming into town. To sign on at the dole."

My heart flipped a beat and I felt a funny little twinge in the stomach. Here she was, four years later and still signing on at the dole.

The half-hour wait was terrible. I rode from Exmouth to Budleigh Salterton. It was a lovely October day and I could have enjoyed it more but for the thought of Linda signing on at the dole. That really was disappointing.

Life as a teacher holds a few magic moments and it was one when Linda and I met outside the post office.

"Come on, then. What's all this dole business?" I asked her.

"Oh, that is just a temporary thing. I'll tell you about it over a drink."

According to her she had not liked our mathematics lessons all that much. "Well, it was different, David," she said, "talking about fractions in a pub. And I understood what we were doing." Then she knocked me down after building me up with, "And I could, for the first time, see what we were doing. But I could never give too much of my time to it. Some of the stuff we did, well, why should anyone bother. No, I am sorry; I felt that I knew what we were doing, felt more confident about doing mathematics, but it would never be my choice."

Then in a magic few seconds Linda taught me a beautiful lesson about mathematics education and about myself.

"There was a big lesson I learnt from our sessions," she continued. "Obviously you had spent a lot of time and energy on maths; it had to be important to you. You made me realise that I had to find something to put my energy into. I knew from what we did it was not going to be maths. But I had to find it. One day I sat on the train going into Exeter. A girl sitting opposite me was reading the *Nursing Times*. I looked at her and thought, 'that's it'. I went to the hospital and signed on a two year training course."

She continued her story, telling me of life at the hospital, the long hours, the parties, dancing with young doctors, the strain of continually being among the sick, the words of thankyou from a patient, watching someone die and everything that makes up a nurse's life. Linda was very wrapped up in the whole package. She had recently finished her training and in a few months time she was due to start on a new diploma course at one of London's teaching hospitals. "It's a course where you are trained to look after little kids who are dying of cancer, leukaemia and such like. It is ever so difficult to get on it. I have had a bit of luck."

Maybe Linda had had a bit of luck getting on the course, but she had also had a lot of determination. That was obvious from the look on her face.

I had never before considered the complexity of handling little children who only have a few weeks of life left. Tears formed in my eyes as Linda calmly remarked, "David, they are going to die and I can't change that. But I can't ignore them and can try to give them something in whatever time they have left."

Later, after the pub had shut, we sat in a coffee house. When Linda went to the counter to fetch a second helping an old lady at the next table said to me, "Isn't she a lovely girl?" I could offer no response.

Quite recently I presented a so-called paper on mathematics education at a university. There are many things in mathematics education which get a continual airing. There is curriculum content, learning theories, teaching methodology, types of assessment and many, many more. As I tried to explain in my own seminar, when we truly examine the mathematical education imperative, or what mathematics education is all about, there can be only one solution. It is about producing more Lindas.

[1]   See R. Harding and D. Kent in *ATM Supplement* No. 17.

INVERSION by Robert Dixon

ASSOCIATION OF TEACHERS OF MATHEMATICS

Inversion, taken from the ATM black & white poster set 'Robert Dixon Posters', by Robert Dixon.

SECTION 2: MATHEMATICS IN THE CLASSROOM

In "Content with Process" Mike Ollerton takes the view that MA1 is the heart of Mathematics and MA2- the body. He uses the non-statutory guidance of the National Curriculum to explore ways of ensuring that mathematical processes are This is followed by Dave Hewitt's article on mathematical investigations in which he encourages teachers not to fall into the trap of reducing all such activities to the single mathematical skill of "spotting number patterns", but to encourage students to delve more deeply into situations and see the breadth and depth of the Mathematics available. Dave's strongly held views are tempered slightly by Paul Andrews' plea that teachers' explorations into ways of engaging students in mathematical exploration must be encouraged even if the explorations are limited initially.

Tony Brown's "Active Learning within investigational tasks" explores ways of structuring investigational tasks so that the teacher becomes more conscious of how children come to understand mathematical phenomena. He introduces the terms metonymy (applying to the internal relations of Mathematics) and metaphor (seeing connections between pieces of mathematics and situations in the real World) as a means of helping in this process.

The next set of articles are about language and Mathematics. In the first of these, Susan Pirie and Rolph Schwarzenberger explore ways of categorising students' discussion amongst themselves as they explore a mathematical situation. They do however advise caution in drawing conclusions about student understanding of concepts from conversation transcripts. In the next article we move to Australia to explore ideas about the use of stories to stimulate mathematical activity. Rachel Griffiths and Margaret Clyne who are based in Victoria illustrate ways in which stories can be used to provide a context, provide a model, pose a problem, stimulate interest and illustrate a concept within Mathematics. There then follow two sections from the ATM booklet "Language and Mathematics" which seek to explore why it is that, whilst language and Mathematics are closely related symbol systems, language is acquired naturally by almost everyone, but Mathematics is acquired very slowly and often painfully by a small minority.

The section is completed by a series of articles, which contain many ideas for mathematical activity . We start in America with Marion Walter who has a wonderful capacity to see Mathematics in almost anything and who manages to encourage others to believe that they can too! "What would happen if......" is one of Marion's favourite questions. This is followed by some Marion-inspired activities from Jane Anderson in Cornwall. They are based on the theme of SQUARES and and suggest a range of activities for students aged 5-19. We end with a visit to the "100 Square" in which Bob Vertes explores a number of activities on which students of all ages could work.

CONTENT WITH PROCESS?

What is the relationship between Ma1 and the other attainment targets?
Mike Ollerton dips into Non-statutory Guidance to find out.

One of the main concerns I hear about from teachers, and particularly initial teacher trainees of mathematics, is that of being able to make Ma1 assessments of the work their students produce. This may be due to an uncertainty about how Ma1 process skills might be incorporated into teaching methodology and consequently transferred into our students' learning. Ma1 skills are often dealt with through 'bolt-on' or separate OET-type tasks and there exists a tension which is the contrast in expectations between students engaging with mathematics through process skills, yet being tested in narrow ways exclusively on content skills. I want to put forward a model which causes Ma1 skills to permeate the mathematics curriculum and at the same time provides opportunities for children to 'learn the basics'.

I believe that National Curriculum Non-Statutory Guidance provides us with not just a great deal to think about, but also some strategies by which we can develop our teaching approaches so that Ma1 becomes an integrated part of learning. In the second part of this article I offer some strategies for possible ways forward.

Extracts from and interpretations of NSG

Activities should enable pupils to communicate their mathematics
Pupils need to:
understand what needs to be done in broad terms; ...
debate possible courses of action with others; ...
present and explain results to other pupils, teachers and other adults;
make a report; ...
discuss the implications and accuracy of the conclusions reached; (B para. 5.12)

There is an image here of students being able to create and explain their mathematics. Teachers need to find strategies to make this possible. This can be achieved through a mixture of short and extended tasks. There are implications about encouraging students to share their mathematics with people other than a teacher, and parents or guardians might be encouraged to play such a role.

Activities should enable pupils to develop their personal qualities
The personal qualities which pupils need to develop include:
motivation and preparedness to tackle the unfamiliar and unknown – willingness to 'have a go';
flexibility and creative thinking in overcoming difficulties and developing new approaches;
perseverance, reliability and accuracy in working through sequences of stages in an extended task;
willingness to check, monitor and control their own work;
independence of thought and action as well as the ability to co-operate within a group;
systematic work habits. (B para. 5.13)

This demands a shift of responsibility, which is clearly not going to be achieved by teachers indicating that 'it is now up to each of you (the students) to develop further without help from me'! There must however be a whole multitude of opportunities for the students to be able to accept ever more responsibility for their learning if they are going to be able to develop such personal qualities and independence. Letting go is one of the hardest things for teachers (and parents) to do. However there must be times when learners can leave the (mathematical) nest and begin to make their own way. There are vital issues regarding teacher intervention that we need to consider and engage with, such as when it is useful to stand back and when it is useful to be didactic. Having to make instant, 'thinking on our feet' decisions is a central role for teachers. It is the making of such professional judgements that causes teaching to be draining and frustrating as well as energising and fulfilling. Of course, we are not always available to provide help and advice, and in large classes such help cannot always be immediate and so other strategies need to be considered. Once such strategy might be to ask another student in the class to take on a teaching role for a few minutes and talk to another member of the class. Group work will obviously make it easier for students to gain help

MT145 DECEMBER 1993

and support in this way. As teachers, we all know that moment when in the course of explaining something, we have come to understand in greater depth what a certain concept really means. This also happens when one student helps another to reach an understanding about an idea.

Activities should enable pupils to develop a positive attitude to mathematics

Attitudes to foster and encourage include:
fascination with the subject;
confidence in an ability to do mathematics at an appropriate level. (B para 5.14)

Here we are encouraged to put our students in touch with the beauty and power of mathematics as a creative and imaginative subject. The final point is I believe crucial and needs to be considered both from the learner's as well as the teacher's point of view. To this end it is important that students are provided with opportunities to develop the task they are offered to whatever level they are capable. This is different to teachers predetermining that, within the same topic area or context, students of different 'abilities' need to be given different tasks to do at the beginning of a topic or piece of work.

It is not the intention of the National Curriculum to produce a narrow mechanistic approach to learning or teaching of mathematics through a rigid interpretation of the system of levels within the programmes of study. Indeed, the programmes of study which relate to using and applying mathematics, with the requirement for pupils to make choices and to work across the other elements of the programmes of study, provide a powerful disincentive to this narrow approach being adopted. (B para 7.6)

This paragraph recognises that effective learning of mathematics is not achieved by the use of a fragmented, exercise-by-exercise approach. Consequently, there is a need to construct a framework or a curriculum map, that helps teachers to decide how else the narrower skills can be taught and to offer students tasks that have the potential for developing such skills. In this way students learn to apply skills within broader problems and wider contexts than can be achieved by working through exercises from text books.

The teacher's job is to organise and provide the sort of experiences which enable pupils to construct and develop their own understanding rather than simply communicate the ways in which they themselves understand the subject (C para 2.2)

This is essentially about the important process skills through which a more controlled and deeper understanding of traditional mathematical concepts can be achieved. Showing students how to be in control of their mathematics in order to achieve ownership, rather than borrowing a set of skills which are on temporary loan from their teacher, is central to effective learning.

The statements of attainment within Ma1 contain the objectives for three strands of mathematical activity:
Using mathematics
Communicating mathematics
Developing ideas of argument and proof
*It is through engaging in these activities that pupils will encounter the real power of mathematics. They are at the **heart** of mathematics, and should underpin pupils' work across all the areas of mathematics in the programmes of study at every stage.*

This is a strong encouragement for teachers to enable students to learn the broader process skills in all the tasks they do. The occasional Ma1 tasks, bolt-on investigations or 'open extended tasks' done in controlled conditions, do not seem to fit with the intentions behind this statement. At the 1993 ATM Lancaster conference I was fortunate enough to have discussions with Peter Lacey (Professional Officer for Mathematics, SCAA) about this article and he provided the following metaphor: Ma1 is the heart of mathematics and Ma2-5 is the body. The body is easier to see but requires the heart in order to exist. With regard to assessment, the health of the heart can be determined by looking at aspects of the body. Peter also provided the following diagrams as a means for considering the interdependence of Ma2-5 with Ma1.

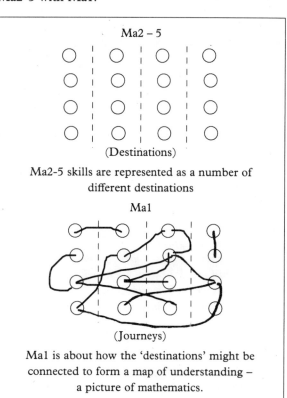

Ma2 – 5

(Destinations)

Ma2-5 skills are represented as a number of different destinations

Ma1

(Journeys)

Ma1 is about how the 'destinations' might be connected to form a map of understanding – a picture of mathematics.

Here the Ma2-5 destinations are connected by Ma1 journeys. The more journeys that are made, the more complete the map of understood mathematics becomes.

Applying mathematics to 'real life' problems

It is characteristic of 'real life' problems that they frequently do not have unique solutions; they require the selection and use of a wide range of mathematics. Applying mathematics to real problems does not come naturally or easily to many pupils, even when their grasp of the relevant knowledge and skills is sound. For this reason, pupils at all stages need to have experience of tackling 'real life' problems as an integral part of their experience of mathematics. (D para. 2.2)

A key issue therefore is transferring skills across different contexts and this is about finding ways of enabling students to gain a depth of understanding about which skills to use and when to apply them. This is different to being able to use the narrow skills that exist on page 57, exercise A, questions 1 to ... There is a further important idea within this statement which is about finding problems that cause students to learn about and acquire new skills. This is different to the strategy of teaching students a set of separate skills and then providing them with broader (convenient) problems that draw upon the taught skills. The dilemma for teachers is one of feeling that certain skills are demanded by a certain problem, and that before the students will be able to 'solve' the bigger problem they will need to be taught relevant skills. I am aware, for instance, that some teachers, before giving the *Octagon loops* task, feel they must first do some work on sequences and functions. If ever tails wagged dogs! There is also a built-in assumption that students can only perform tasks if we have first taught them. I believe that many of our students are more intelligent than we often give them credit for. Eventualities such as students applying: intuition; ideas despite my teaching; and concepts that I cannot possibly know that they already have; are all possibilities that I need to take into account.

Pupils exploring and investigating within mathematics itself

Mathematics provides a way of viewing and making sense of the real world. It is also a way of creating new imaginative worlds to explore ... this aspect of mathematics which encourages pupils to explore and explain the structure, patterns and relationships within mathematics is an important factor in enabling them to recognise and utilise the power of mathematics in solving problems and to develop their own mathematical thinking. (D para. 2.3)

Finding suitable tasks and planning how to introduce them to pupils needs to become central to teachers' preparation. Having a variety of such tasks to draw upon and recognising what works for them as teachers and is effective for their students' learning takes time and experience to develop.

The National Curriculum requires all schools to address this issue, and develop a teaching and learning approach in which the uses and applications of mathematics permeate and influence all work in mathematics. This is a major undertaking for schools, and perhaps the single and most significant challenge for the teaching of mathematics required by the National Curriculum in its aim of raising standards for all pupils. Schools must ensure that schemes of work offer sufficient opportunities for activities which specifically address the issue of using and applying mathematics. In addition, schools must carefully review their approach to teaching knowledge and skills to ensure that aspects of using, applying and investigating are integrated and embedded into the ways in which mathematics is taught and learned. (D para. 3.2)

This is probably the most powerful statement in NSG and encourages departments to look at current practice, to review methodology and to consider alternative approaches to the teaching and learning of mathematics. An example of an alternative approach is provided in the following section.

Some strategies for incorporating Ma1 skills into teaching and learning.

In order for students to be able to demonstrate that they are using and applying mathematics they need to be provided with opportunities to do so. This essentially requires a shift from teachers seeing their role exclusively as givers of knowledge, where learners are requested to carry out a collection of examples in order to show that they have received the knowledge, to other modes of teaching and learning.

Unfortunately, a teacher's role has become muddled when it comes to providing students with certain types of input. Thus, tasks such as *Octagon loops*, which have become established as the 'type' of task that teachers can use with the expectation that students responses can be matched against Ma1, are all too often seen exclusively as assessment instruments; teachers feel they cannot give too much help or tell individual students anything. This is clearly a nonsense because the teacher's role has become subverted from that of helper to one of observer and assessor. I believe that Octagon loops should be seen as just another teaching task through which assessment might be made. There can therefore be a wide range of possible teacher interventions, ranging from standing back and observing to being didactic and telling. The type of intervention, including the decision not to intervene, must be left to the teacher's professional judgement: it will be whatever best suits the learner and enhances understanding and progress.

There are many similar starter tasks which could be deemed 'open' and these can be found in resource books such as: *Points of departure* (Books 1, 2, 3 and 4) – published by ATM and *Starting points* (Banwell, Saunders and Tahta) published by Tarquin. These publications alone will provide well over 200 potential ideas for use in the classroom.

Determining starting points

I am more interested in constructing tasks that address a range of the Ma2-5 content skills, but are driven by Ma1 process skills. One of the more negative teaching situations that has existed within my own classroom has been where I have started working from a particular point, and, realising that many of the students had little idea of what I was talking about, I have had to take a backward step. This possibly continued until I thought I had found a point where most of the class were comfortable. By such time I had already turned off some of the less confident students and I had wasted the time of the more confident ones who were comfortable with my original starting point. Thus, I had provided a negative set of experiences for just about everyone! An alternative to this is for me at the planning stage to decide what high level area of content I might wish students to end up at and then decide what easy point I might begin from. To do this I can go further back until I feel comfortable that everyone in the group will be able to understand the beginning point. I can then offer a problem for exploration based on this beginning point. The most capable mathematicians will be expected to develop the starting problem and quickly move their thinking into more complex ideas, that I can provide further input for. This approach caters for a wide range of responses and provides learners with a positive, constructive approach to their mathematics.

For example, I want my students to explore volume. In order for them to find ways of taking control of concepts of volume, I use my knowledge of volume to find real opportunities to release my control. This must happen in a planned way. One teaching strategy is to set up a relatively simple problem and then set a problem that causes the students to explore what is happening and look for a set of results. My simple problems could be to give each pair of students 24 Multilink cubes and ask them to find all the different cuboids that can be made using all 24 cubes. I can then ask the class to calculate the surface area of each cuboid and from here I have a choice of possible ways forward.

1. Volume = 60cm^3

- explore different cuboids and surface area for V = 60;
- sketch some results on isometric paper;
- calculate the surface area of each – minimum S.A.;
- use the calculating procedure to construct a formula for surface area;
- what would the minimum surface area be if I allow non-integer lengths;
- change the shape to a cylinder, what different r and h gives V = 60;
- the above problem can be solved using a programmable calculator;
- what values of r and h give minimum surface area.

2. Volume $\leqslant$ 30 cm^3

- explore all the different cuboids that can be made for V $\leqslant$ 30; this could possibly be done in groups;
- find values which give only one cuboid (prime values);
- collect together surface areas of all the shapes that have dimensions 1 by 1 by n, 1 by 2 by n. Number patterns and the resulting formulæ can then be sought and related back to the original situation;
- collect together sets of dimensions which have the same surface area.

An important feature is that at various stages students are being encouraged to explore the ideas and to construct a meaning from the results they achieve. Again the teacher will offer varying amounts and types of input, either with individuals, with small groups or with the whole class. Thus differentiation occurs by outcome and by task.

Either approach can lead to students writing up their findings with explanations, diagrams, tables of results, graphs, formulæ and possibly a final section where students are asked to reflect upon their learning. This could be a list of ideas that they: know at the end of the project; had been reminded of; didn't know or had forgotten prior to starting the work. This can then link in with a student's record of achievement.

Mike Ollerton has a joint appointment at Orleton Park School, Keele University and St Martin's, Lancaster.

TRAIN SPOTTERS' PARADISE

Mathematical exploration often focuses on looking at numerical results, finding patterns and generalising. **Dave Hewitt** suggests that there might be more to mathematics than this.

I have been in many classrooms where children have been encouraged to use their intelligence and creativity to find some mathematical properties. Children have been asked to look at particular situations and encouraged to find connections, make conjectures and test to see if those conjectures are correct. They are encouraged to make generalisations and to express those in algebraic form. In such lessons, I am impressed by how much children are able to discover for themselves and how well they can articulate their findings. There is an atmosphere of involvement in mathematics, children are being challenged and are expressing a sense of achievement in what they are doing. Quite often they continue working on their problem at home and involve their parents. They may arrive the following day eager to share new things the family have discovered. Such times have so many ingredients of lovely mathematics lessons. Yet I feel saddened rather than joyful.

I will mention five such lessons:

In one lesson, children are asked to draw a number of networks and to see whether they can be traversed without taking the pen off the paper or going over any line twice. After a while the teacher asks them to draw up a table giving the number of nodes, how many are odd and even, the number of arcs, the number of regions, and whether the network is traversable or not. The challenge for the children is to look at the table and see whether they can see any patterns.

In another lesson, children are looking at the number of matches required to make a square such as:

The children draw different sized squares and collect their results in a table. A number of patterns are found and many children articulate rules.

In a third lesson, the class have been asked to draw a number of circles with different radii. Using some string, they measure the circumferences of each circle and make a table of the radius, diameter and circumference. The children are asked to try to find a connection between the radius and the circumference.

The fourth lesson involves choosing a number, say 68, reversing the digits to get 86, and adding the two numbers together:

$$68 + 86 = 154$$

If the answer is a palindrome, stop. Otherwise, repeat the process with the answer:

$$154 + 451 = 605$$

$$605 + 506 = 1111$$

In the case of 68, it took three iterations to arrive at a palindrome. Thus the number 68 is called a *level 3* number. The class are divided into groups and, between them, are asked to find out what level are all the numbers from 1 to 100. They are asked to collect the results in a table and to look for patterns.

In the fifth lesson, children are listing the different outcomes that are possible if they throw 1, 2, 3, ... coins. They are asked to collate their results in the following table:

	Number of ways of getting:					
	0 heads	1 head	2 heads	3 heads	4 heads	. . .
1 coin						
2 coins						
3 coins						
4 coins						
. . .						

Then they are told to look for patterns and predict how the table would continue.

Despite the fact that in each of these lessons children were well motivated and involved in mathematics, I am saddened because the children ended up doing a similar activity irrespective of the initial mathematical situation.

Is the diversity and richness of the mathematics curriculum being reduced to a series of spotting number patterns from tables?

Whatever the initial mathematical situation, once the numbers are collected into a table, a separate activity begins to find patterns in the numbers. Their attention is with the numbers and is thus taken away from the original situation. After a period of time, some children have difficulty reminding themselves where all the numbers came from. I suggest that for many children, what they find out about the numbers remains exactly that; it does not mean they have learnt anything about the original mathematical situation, only about sets of numbers in a table.

Children can find many patterns in their table, even if they have made some errors in the entries. They may find all sorts of rules, none of which apply to the original situation but then some children have long ago turned their attention away from that. Spotting patterns in the numbers becomes an activity in its own right and not a means through which insights are gained into the original mathematical situation.

Networks may come under a heading of topology; the square of matches is essentially a geometric situation; circumference of circles may come under a heading of measures or geometry; palindromic numbers within number theory; and the coins within combinatorics or probability. These initial situations span a broad cross section of mathematical areas and yet I argue the case that each of these lessons were really under the same heading of spotting number patterns since that is what the pupils ended up attending to.

In all these lessons, the children were doing several particular examples and collecting results from these. I presume the structure of collecting results in a table offers the possibility of making general statements about these results. The trouble is that the general statements are statements about the results rather than the mathematical situation from which they came. The existence of the table places value on collecting several results rather than looking in any depth at a particular one. More might be learnt about the original mathematics if one particular situation was looked at in depth, rather than rushing through several in order to collect results.

If I consider the following network:

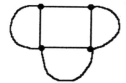

I can learn that it is traversable. However, instead of rushing on to consider another network, I could explore this one further. Are there other ways I can traverse this network? What if I keep the same starting node, where can I finish up? What if I try starting from the other nodes? How often do I visit each node? What would change if I rubbed out one of the arcs? Does it matter which one?...

If I take some matches and start putting them down so as to make this square:

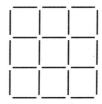

In what order do I place them down? Do I repeat certain patterns of matches? How often? Is there a stage when I am half way through? Can I look at the final square and imagine half the matches a different colour? What about a third? A quarter? How many horizontal lines are there? How many vertical? Why? Can I see the square one size less within this square? What if I paint the extra ones blue? Can I see the square two sizes less within this one?...

Imagine taking the diameter line of this circle:

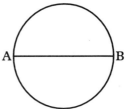

and picking up a copy of it leaving the circle with the original diameter staying where it is. Imagine I could bend the diameter although I cannot alter its length. I try to bend it so that it curves round the circumference of the circle. If this bent diameter starts as A, how far round the circumference do I think it will go? Will it go as far as B? Suppose I put a mark where it has got to and continue with another copy of the diameter, how far now? How many diameters would I need to go once round the circumference and return to A? What if I did a drawing of this and put it in a photocopier which reduces the size?...

If I consider adding together 68 and its reverse 86:

What numbers appear in each of the columns? If there are the same numbers, why do I get the same number in each column of the answer? When will I get the same? What effect does the 'carry' have? How could I change the numbers so that I do not get a carry? If I stick with the original number being 2-digit, when do I get 2-digit answers? 3-digit answers? Can I get 4-digit answers? If the answer is 3-digit, what digits could I get in the hundred column?...

Let me consider the situation where I have exactly one head with a number of coins. If I know how many ways there are of getting one head with two coins, do I have the same number again when I introduce a third coin which happens to be tails? If the third coin was a head, what would the other two coins have had to be? What if I consider a larger number of coins and introduce one more coin which is a tail? A head?...

There is so much mathematical richness that can be gained by looking at a particular situation in some depth rather than looking at it superficially in order to get a result for a table and then rushing on to the next example. By staying with the particular situation, I can learn about the mathematics inherent in it rather than learning about numbers in a table. I practise and develop different abilities rather than practising and developing the one ability of spotting number patterns. I see geometry as geometry, combinatorics as combinatorics rather than everything as spotting number patterns. I am being asked to be creative and adaptable in different situations, to see that different situations require different questions to work on them.

Train spotters go in search of trains and collect numbers. At the end of the day, they are left with numbers... not a train in sight.

Dave Hewitt teaches in the School of Education at the University of Birmingham.

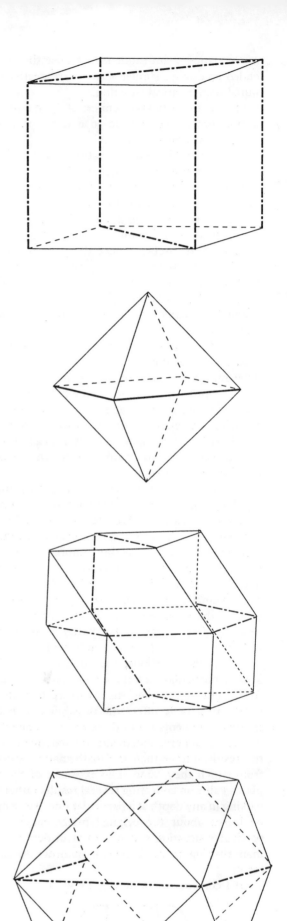

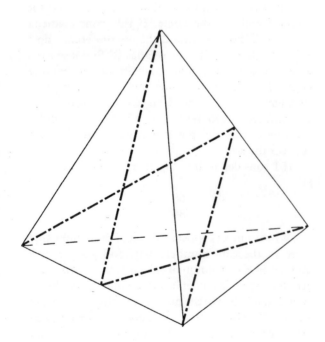

TRAIN SPOTTERS HAVE FEELINGS TOO

In Train Spotter Paradise it was suggested that number patterns inappropriately dominate the exploration of mathematics in schools. Here **Paul Andrews** responds.

I have been contemplating Dave Hewitt's article Train Spotters' Paradise (MT140) for several weeks now and, despite finding much of interest, feel discomforted by some of the issues he raises. Certainly he struck an emotional chord – and this may have been his intention – which has in some ways added to my discomfiture. Indeed, I have begun a response on at least four occasions and each time I have ground to a halt. I considered starting with a paragraph of praise for those about to be condemned but decided that such an approach lacked intellectual honesty. I thought about writing from the heart – in the sense that I just let my feelings flow – but this too seemed to have a dubious integrity in that the reader would never really know whether or not such outpourings were genuine.

Indeed, this dilemma of interpretation must be at the heart of all writing and, it seems to me, makes Dave's article all the more interesting. On the one hand, it is possible that he has taken an issue and written a piece intending to provoke a reaction to it. If this was the case then he has succeeded in his provocation if only because he has forced me to respond.

On the other hand one must accept the article at face value. In which case it must reflect what Dave believes about the nature of mathematics and its teaching – and I suspect that I am in sympathy with him over much of what he says – but I do not believe that the manner in which he has put his argument across is helpful to the people he has implicitly criticised.

So what was it that Dave says that left me wanting to respond? In short, his article suggests that too many teachers focus too closely on number patterns in their teaching. He believes, irrespective it seems to me of the context of an investigation, that opportunities for children to explore worthwhile mathematics are lost as teachers encourage the quest for a generalised number pattern. In support of his argument he offers criticism of several activities he has observed. These include an investigation of matchstick shapes, some work on the traversability of networks and a guided discovery of the relation between the diameter and circumference of a circle.

In some respects I am inclined to agree with him. He suggests that 'spotting patterns in the numbers becomes an activity in its own right and not a means through which insights are gained into the original mathematical situation.' He goes on to say that once children have collected together their results and looked for the general statements '... the general statements are statements about the results rather than the mathematical situation from which they came.' I believe the message he is trying to put across is an important one and find little in with which to disagree. So just why did I feel the need to reply?

One has only to look at almost any school text or commercial scheme to see the limited expectations placed on most children and the turgid, repetitive and undemanding diet offered them. What concerns me is that Dave's criticism seems directed at those teachers or student teachers who are trying to do something about it. If we are to take his first paragraph at face value then the lessons he saw were more about children learning and doing mathematics than the transmission of some piece of content – whether by SMP, KMP, STP or any other of the ubiquitous schemes that lead to a lack of progress in mathematics – which makes me wonder whether he has decided who the enemy really is.

Worse, and this is a potential disaster, if those people who are attempting change – despite all the traumas of the last five years – see their efforts undermined by such criticism, is there not a risk that they might revert to the didactic modes of operation that characterise all that is disappointing in mathematics education, but which both pupils and teachers seem to find comfortable and unthreatening?

In the first instance, I wonder if Dave is the most appropriate person to criticise, in the way that he did, the content of the lessons he described. I do not know him well, I have met him once, over lunch at a conference, but I do know that he is held in high esteem by many of my colleagues. They speak of his ability for making mathematics accessible to both children and adults. They talk of lessons they have seen him teach – both 'in the flesh' and on video and have commented admiringly on the powerful way in which he works and on the high expectations he sets both himself and his students. Yet, by criticising in the way he did the lessons he observed, perhaps he fails to recognise his good fortune. He is where he is because of the qualities he possesses, because he can see new avenues where others cannot. He is the exceptional teacher. His is the extraordinary talent.

In all aspects of human endeavour there will be some more talented than others. Without those players ranked two hundredth in the world there would be no competitions for Seles or Graf to win. Christie would never have achieved Olympic gold if other, less fortunate, athletes had decided not to compete and allow him the chance to shine. Such people recognise their good fortune and know that success is fragile and fickle. They recognise, through work with their coaches and others in their sport, that changes are not brought about by criticism but by encouragement, reasoned argument and gentle persuasion. Teachers are no different and it should come as no surprise that many, even those attempting change in their practices, fail to match up to the expectations of those regarded by their peers as among the eminent of the profession. Whether in teaching or sport, those who are committed give of their best. They train as best they can within the constraints of individual circumstances. They practise their skills and try to improve on their performance. In short, if the mud is to be thrown then surely it should be aimed at those who offer nothing of value rather than at those who commit themselves to some form of growth, change, uncertainty.

I wonder if those of us who consider ourselves to be confident and experienced in our work remember what we were like as beginners? Were our skills as sophisticated as they are now? Were we always able to see potentially productive lines of enquiry? Was the underlying structure of a situation of paramount concern for us from the moment we walked into a classroom or was it something that developed with time as we gained both confidence and competence? Were our expectations always as high as they are now or did they grow as we grew? Are expectations related to experience? Such questions are important because the answers help shape our growth as teachers. It would be unrealistic, indeed unfair, to expect student teachers or, for that matter, anyone contemplating change to have the insights I think I have.

There is another related issue to do with opportunity. Increasingly it seems that schools are gearing their work to their perceptions of the needs of the National Curriculum. They are using arguments about curriculum coverage and the difficulties of assessment as weapons against any form of innovative practice. 'We must guarantee pupils' entitlement to such and such' seems to be the watchword. It will be a brave teacher who shows a metaphorical two fingers to the establishment and goes her own way. In other words, there are pressures on teachers to conform to mediocrity and those who are trying not to do so need our support rather than our condemnation.

It seems to me that Dave's article fails to address several important points. We do not know what the intentions of the different teachers were for their different lessons. We do not know anything about the skills of either the children or the teachers involved. We know nothing about the pupils' or the teachers' previous experience. We know nothing of the context of the lessons discussed. If, for example, a class has been exposed only to a narrow, well-defined and didactic approach to mathematics then the risks associated with too radical a change to what is offered them are well known and too great for many teachers to attempt. If a child's experience of learning mathematics was as a recipient of transmitted information then it is unrealistic to expect that same child to accept immediately the notion of independent investigation and discovery. In other words, teachers behave according to the context in which they work, and change, if it is to be achieved, usually comes slowly.

The thrust of Dave's article, based around five lessons, seems to be that effective learning is undermined through too great an emphasis on work with number patterns. This may be right, but I sense that there are other issues involved that need raising. The article is only in part about five teachers and their lessons. It is also about teachers' perceptions of the nature of mathematics. I confess that I find number patterns fascinating and see them as being at the very heart of the subject. I do not, for example, necessarily see things spatially. I recall a series of lessons I shared recently with a group of three post-graduate students. They had set up an activity whereby an average-ability group of year nine pupils had been investigating the angle sums of polygons. This led, quite naturally, to their being asked to find the interior angle of each polygon and to graph it against n, the number of sides of the polygon. We chatted to individual pupils about their interpretations of the graph, the notion of a limit and about what was happening in this instance. They were then asked to draw on the same axes a second graph of the exterior angle against n and to comment on what they noticed. It wasn't exceptionally exciting but at least we felt that the pupils had been meaningfully engaged in relevant geometric discovery. That evening I decided, whilst pottering with some ideas for the next lesson, to investigate the ratio of interior angle to exterior. I didn't recall ever having considered it before and I didn't know for certain what the result would be. I knew, from experience, that the result, though not necessarily of great significance, might be of interest. I expressed each ratio as a decimal and was stunned by the simple elegance of what I had found. It is likely that readers will be familiar with it but that doesn't detract from my having found it for myself.

The point here is that I was able to discover what is, in essence, a geometric result through working with number patterns. I find it satisfying that my ability to work with numbers allowed me to do it, but am still intrigued by the geometrical implications. I didn't lose sight of the source of the number pattern – although I accept that this may be related to my being able to hold one notion in my head whilst working on another. Nor do I feel that my sense of wonder over a geometrical notion has been diminished through my discovering it in the way that I did. I don't think I would have found it in any other way. My way reflects how I view things. It is indicative of my particular interest in mathematics and of my idiosyncrasies. I accept that I may be poor with mental images. That doesn't mean to say that I never try to use them or undervalue their power. It does mean that working with number patterns may allow me access to a situation that otherwise might remain closed. When I was at university I had the usual first-year course in linear algebra, which was given by a man well-known for his enthusiasm for steam trains. At the time, as indeed I think I still am, I was painfully shy in large gatherings and had never offered a response to any of the questions asked of us. One day he said he wanted to demonstrate a point with a numerical example. He wrote a number on the board – 4472 – and asked the hundred plus people present if anyone knew of its significance. For the first time in my life I felt sufficiently confident in my public ability to get a question right. "The Flying Scotsman", I blurted. "Splendid", he replied. It was the only question I answered all year.

Paul Andrews teaches at the Manchester Metropolitan University.

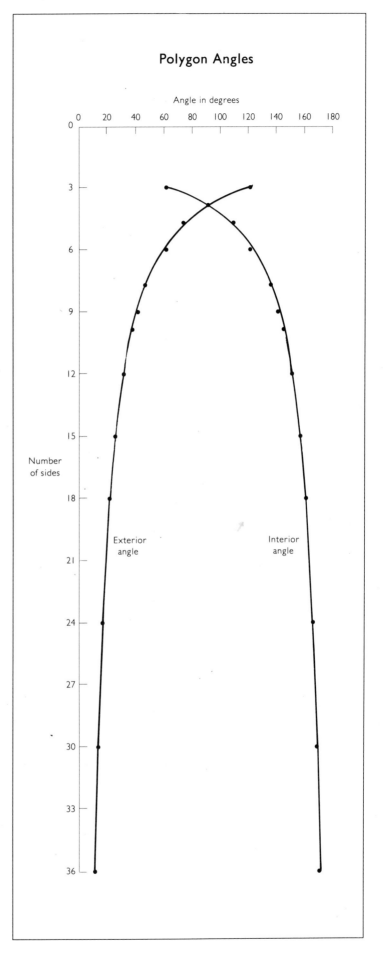

ACTIVE LEARNING WITHIN INVESTIGATIONAL TASKS

Tony Brown

Investigational tasks within a teaching schedule enable children to engage in personal mathematising and offer an alternative to tasks that require a pre-defined response. Whilst some teachers prefer to allow children to proceed through such work with minimal guidance, others express disillusionment at the result of such work appearing to lack purpose or any clear result. Here I wish to demonstrate how some recent theories of learning might enable the teacher to change the emphasis in their structuring of such tasks to be more conscious of how children come to understand mathematical phenomena.

Valerie Walkerdine has challenged many of the ideas underlying current practice in schools and in particular has questioned the strong faith many teachers have in 'doing' practical tasks as a sufficient requirement for understanding mathematical phenomena. She suggests that children come to know mathematics through participation in mathematical activity with others, including the teacher, and in this way the children learn to talk about and engage in mathematics. Towards describing this she has introduced the terms *metonymy* and *metaphor* into the vocabulary of mathematics education. Metonymy might be seen as applying to the internal relations of mathematics, that is, the theoretical side of mathematics, before any attempt is made to relate this to any practical purpose. One aspect of metaphor is seeing the connection between pieces of mathematics and situations in the real world they seem to model. For example, within mathematics we can talk of rectangles, squares and triangles and develop theories about them and these seem to work when such shapes are found in the real world. But it is also helpful to see metaphoric association more generally as recognising similar mathematical phenomena in different contexts.

Walkerdine suggests that learning takes places as we move between metaphoric and metonymic attentions. Initially she sees learning being tied to familiar situations with the children engaging in what she calls the social practices arising in learning situations. This she sees as being not so much concerned with ideas developing in the mind; rather the children are learning how to talk about and engage in mathematical activity. However, for children to be capable of more abstract thinking, she suggests they need to be able to suppress their attention to metaphoric associations concentrating instead on the internal relations of the mathematics ie the metonymic associations. I feel it may be helpful for me to describe an investigational task that I initiated with a class of ten year olds, over several lessons, to illustrate how I see these ideas having very practical relevance.

I selected a task called *Paths* described in the Manchester Polytechnic booklet, *Leap To It*, compiled by Gillian Hatch. I varied it slightly to consider rectangular gardens with a metre wide path around the perimeter of a lawn.

To commence this work I provided everyone in the class with plastic Polydron squares. I held up a 'lawn' I had made and showed them how it looked after I had surrounded it with squares making a 'path'. I followed this by making other lawns myself and asking the children to show me what each would look like after being surrounded by a path. I had arranged the children on tables of four and had only provided enough pieces for a group effort and so some sort of negotiation between them was necessary. After a few examples I became fairly confident that each group were able to respond to a Polydron lawn made by me with a similar sized lawn surrounded by a path. At this point I asked them to make different ones.

In describing such gardens to the children and them representing these in plastic models, they are perhaps making a metaphorical association between the image of such gardens in their minds and the plastic models before them. However, they were in some sense suppressing this metaphoric association as they focused on exploring different sorts of plastic models obeying the given rules. Here they are focusing on the internal (metonymic) rules and relations pertaining to plastic models. There is also another sort of representation evident in this activity. It is in the way they talk about the model. In saying something like *This is the lawn and it's got a path around it* they are metaphorically associating the image in their mind or a plastic model with some words.

When it seemed to me that the children had made a number of plastic models obeying the rules, I suggested they find as many as they could where the lawn was only one square wide. I also handed out cm square paper and suggested they draw all the ones they found after first making them out of Polydron. The various groups proceeded at different rates and I did not suggest drawing until I felt children were beginning to find the Polydron pieces cumbersome. Many children stayed with the plastic models throughout the first lesson whilst others confidently moved into scale drawing representations of them. Many of these children had drawn the gardens with a lawn of width one in order of size. For those who had found them more randomly I suggested that they be re-drawn in order of size.

In initiating this shift to scale drawing representation of gardens, the children are encountering a new metaphorical association. After imagining gardens and then making models of them, they are now employing the representation of scale drawings. As the children explore different sorts of scale drawings attention is moved away from the plastic models. This could be described as the children exploring the metonymic field of scale drawings of gardens.

For children who had completed a number of scale drawings showing gardens with lawns of width one and in order of size, I suggested they tabulated their results in the following format:

Number	1	2	3	4	5	6
Area of Lawn	1	2	3	4	5	6
Area of Path	8	10	12	14	16	18

Most seemed to find this relatively easy although for some it involved going back to the Polydron models or scale drawings to check results. On completing such a table I intervened and asked them what they could say about the table. I received comments such as The lawn is always the same number as the garden *and* The path goes up in twos. *I suggested they write about all of the things they could notice. For most children I suggested that they now try to construct a table of results for gardens with lawns of width two. I suggested they use either Polydron or scale drawings if they wanted to, to help them find the results. Most children seemed to do this but after a few drawings they were normally able to carry on the table without using other representations.*

For more able children I was more demanding in my questioning after they had completed the first table. For example, what could they tell me about the next garden? Or what about the tenth garden? Or, if you know the area of the lawn how do you work out the area of the path? What can you say about the

hundredth garden? What can you say about the nth garden? I attempted to match my level of questioning to the sort of responses I was getting from individual children but in all cases I insisted they wrote everything down in prose.

Here I see two new metaphoric representations being introduced and related by the children to those they had employed already. Firstly, tabulation provides a valuable, perhaps less cumbersome, alternative to plastic models or scale drawings. The children were normally, after some work, able to see the internal (metonymic) workings of these tables without reference to other representations, that is they could continue the tables without reference to models. Secondly, the children were representing their work in writing. To emphasise the importance I see this having I shall reproduce exactly the written work of one child in its entirety. It becomes especially interesting as patterns are described.

Gardens

by Stephen Watkins, Peter Edwards, Andrew McAndrew and Simon Lilley whilst at Downside Middle School, Newport, Isle of Wight

Stephen writes:

We started off with a border with a 4 × 3 area. The path was 1 metre wide. We had to work out the area of the grass in the middle of the square. Peter has done a series of diagrams to show the shape of the garden.

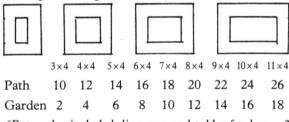

	3×4	4×4	5×4	6×4	7×4	8×4	9×4	10×4	11×4
Path	10	12	14	16	18	20	22	24	26
Garden	2	4	6	8	10	12	14	16	18

(Peter also included diagrams and tables for lawns 3 and 4 metres wide)

We also had to work out the area of the path. Here is the answer for the 4 × 3 square. The answer for the grass is 2. The answer for the path is 10. The next thing we had to do was the 4 × 4 square. The answer for the grass is 4 and the answer for the path is 12. The 4 × 5 square: The answer for the grass is 6, and the answer for the path is 14. The grass area of 4 × 6 is 8 and the area of the path is 16. The grass area for 4 × 7 is 10, and the path area 18. 4 × 8's grass area is 12 and the path area is 20. 4 × 9's grass area is 14 and the path is 22. 4 × 10's grass area is 16 and the path area is 24. 4 × 11's grass area is 18 and the area of the path 26.

It goes up in two's, 2, 4, 6, 8, 10. Also if you look at Peter's diagram for the drawing of the 4 ×, take away 2 squares and you get the starting number for the grass area. The area of the path also goes up in two's. On the 4 × 3 square, the area of the grass and the area of the path, if you take away the grass from the path, you get 8. Do that on the rest of the 4 ×, and you always get 8.

Here is a chart to make it clearer.

Dimension	4×3	4×4	4×5	4×6	4×7	4×8	4×9	4×9	4×10	4×11
Area Grass	2	4	6	8	10	12	14	16	18	20
Area Path	10	12	14	16	18	20	22	24	26	28

Between 2 and 10 is 8, between 4 and 12, 8 between 6 and 14, 8, and so on. The grass area goes up in two's and so does the path area. But in 5 wide, or 5 ×, it does not go in 2 and 2. The area of the path stays the same, but the area of the grass goes up in 3's. Between the grass area and the path area for the 5 × 3 is 9 and for the 5 × 4 it is 8, for the 5 × 5 it is 7, so 5 × 6 would be 6. You can still get the starting number by -2 off the path.

Now onto the 6 ×. The area of the grass is 4 for a 6 × 3 square. To get 4 you do what I have been writing all the time, you -2 off one of the sides of the path.

The 6 × 4 grass area is 8, 6 × 5's area of the grass is 12. 6 × 6's area of the grass is 16. Between those number is 4. On the 4 ×, it was 2 between the numbers off the grass area, then on the 5 ×, it was 3. The 6 × it was 4. The area of the path starts with 14. To get that number and 12 for 5 ×, and 10 for 4 ×. Between the grass and the path area on 4 ×, it is 8, on the 5 ×, the number between is 9. On the 6 × the number between is 10. So on the 7 × which I am not going to do it would be 11.

Next we had to do 10 ×. The grass area started with 8 for the 10 × 3. The 10 × 4 square started with 16. The grass area that is. The grass area for the 10 × 5 is 26. The 10 × 6 grass area is 32. The 10 × 7 is 40. The 10 × 8 grass area is 48. The 10 × 9's grass area is 56. So 10 × 10's grass area is 64. Between those numbers is 8. The path area starts with 22. It still goes up in two's. Between the grass number and the path number is 14 for the 10 × 3. For the 10 × 4, the number between is 8. For the 10 × 5 the number between is 2. It is 6 between.

Now onto the 100's. The area of the grass on the 100 × 3 square is 98. The area for the 100 × 4 on the grass is 196. I don't have to go any higher because you can work it out from these two numbers. The number between them is 98. So, all you've got to do is work out your 98 × table! The path area is still in the 2's. It starts with 202.

Now for the N ×. The N × 3 grass area is N. So is the path area. The N × 4 grass area is N × 2.

The path area is N + 2. So the path area for the rest of the N's would be plussing 2.

Whilst this is the most articulate piece of writing I received, the mathematical work described in it was tackled by many children in the class. This is a fairly unusual and perhaps cumbersome way of presenting mathematical work but I suggest there is great value in the children being able to do this. It seems to me that in translating the work as represented in Polydron models, scale drawings, tables and spoken description into written description, the child will reflect on the rules and relations inherent in each of these metonymic fields. The child is actively moving within a metonymic field in attending to one representation but in the act of writing the child is actively placing different metaphoric representations alongside each other. I feel this active movement between metonymic fields (ie making metaphoric leaps between metonymic fields) underlies children developing an active relationship to the mathematics they are doing.

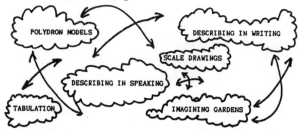

Although most children did not proceed beyond here, some, including those mentioned, were able to extend their results. Many children were able to say how to go about working out the areas in a garden of a given width but for any length. For example, they were able to say something like you add the path at the top to the path at the bottom and then the two sides and then add 4 for the corners. *However, a few explored the possibility of imagining any garden. Andrew was delegated to write the description of this work. I include an extract.*

We had to work out the area of a M by N garden. The area we had to work out was the area of the path and of the garden. N would equal any number while M would any number N doesn't. The path round the garden was always a meter wide no matter how big the garden is. Simon has drawn some diagrams of the gardens.

Area of grass M-2 × N-2
Area of path M × N-M-2 × N-2

This new representation is no more than a shorthand for all the talking and writing that has gone on about gardens obeying these rules but becomes a powerful statement of generality with a whole story around it.

The final stage tackled by some children was concerned with recognising that they had only observed one sort of garden. I here reproduce some diagrams by Simon that indicate possibilities for further extension.

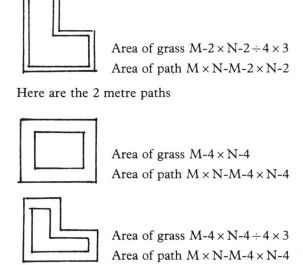

Area of grass $M-2 \times N-2 \div 4 \times 3$
Area of path $M \times N-M-2 \times N-2$

Here are the 2 metre paths

Area of grass $M-4 \times N-4$
Area of path $M \times N-M-4 \times N-4$

Area of grass $M-4 \times N-4 \div 4 \times 3$
Area of path $M \times N-M-4 \times N-4$

The new representation of gardens here extends beyond those that are rectangular with a metre wide path. The metonymic field here suggests an exploration within the symbolic notation together with diagrams that no longer display the discrete properties inherent in the Polydron models. However, it seems to me that for the children involved this notation is supported by the imagery they have explored and developed in earlier work.

In the light of Walkerdine's work, I suggest the children's work in my lesson was following various routines and ways of working which had been established. Tied initially to work with tangible objects such as Polydron, they gain experience of exploring and comparing different metonymic fields. 'Doing' is a side issue here. However, through such experiences they become able to dispense with these relatively concrete representations and work increasingly with more abstract notions.

It seems to me that many investigational tasks have the potential for allowing the children to engage in this active relation between metaphoric and metonymic attentions and so facilitate effective learning. Further, the teacher is in a position to manage the situations the children are working in without undermining the decision-making potential that such tasks have for children. It is evident that the child's work in respect of tasks developed in this way does not fit comfortably into the categories conventionally found in content-oriented syllabi. However, there clearly is 'content' and it seems inappropriate to classify this task as solely 'process-oriented'. A more complementary relation between process and content needs to be recognised along with the notion that the 'content' of mathematics is only ever seen by humans within mathematical activity. ∎

Manchester Polytechnic

References

Valerie Walkerdine *The mastery of reason*, Routledge 1989
Valerie Walkerdine *From context to text: a psychosemiotic approach to abstract thought*, in M Beveridge *Children thinking through language*, Arnold 1982

MATHEMATICAL DISCUSSION –
IT'S WHAT YOU SAY *AND* WHAT YOU DO.

Susan Pirie and Rolph Schwarzenberger

We were excited to read the account in MT121 of the research by a small group of Cheshire/Wirral teachers on categorisation of children's mathematical talk within the primary mathematics classroom. We have been conducting parallel work in secondary schools. Like Derek Smith we have concentrated upon mathematical discussion between the pupils themselves: so-called 'teacher-pupil discussion' rarely fits our definition of mathematical discussion which is given later in this article.

Non-verbal interactions

Like Derek Smith, we have found that mathematical discussion is relatively infrequent and that awareness of the non-verbal interactions is often essential when analysing the talk.

In this extract two pupils are using a function machine:

Z Those?

T No, those.

Z Yeah, it's 12 – put that in the box so you go – right – there's 12 there and you move it across to the 10 and it says 10 minus 12 which is minus 2.

T Yes.

Z And then you do the same for 11 – put 11 there – put it into the machine and the machine comes out there and says: 10 minus 11 ... and the answer would be minus 1.

T And that was...

Z Put 39 there and 10 minus 39 is 29....minus...put a 41 in there and 10 minus 41 is minus 31.

T Yeah, O.K.

Z And you put 8 there, 10 minus 8 equals...

T Thank you.

The discussion is admittedly one-sided, but it depends crucially upon the brief reactions of Tessa as well as upon the understanding of Zara.

From the transcripts the episode might seem insufficiently interactive, but the evidence of the classroom observer and the intonation of Tessa's voice indicate clearly her participation, concentration and comprehension.

This example also illustrates a category of talk which could be considered as the antithesis of Derek Smith's Category 4 – the difference in understanding between the two pupils is being exploited to positive ends rather than used to undermine the more passive participant's confidence. We have found this category of talk – the pupil as teacher – quite common in the classrooms we have observed, although admittedly it does not always qualify as discussion.

Background to our project

Our work, which is still in progress, comes from a slightly different starting point than that of Derek Smith's group, and is a longitudinal case study intended to shed light on the question of whether or not discussion in the mathematics classroom is an aid to understanding.

We were wary of the gap between how we would idealistically like to believe that children learn, and how they actually learn in real classrooms. Therefore, we did not design any experiments, and did not attempt to impose or stimulate any special activities. We observe talk which arises in the course of the day-to-day work taught by four selected teachers in their normal way. However, because we are wanting to study understanding and discussion, we did select for our case studies environments in which we could be sure that mathematical discussion would occur. We approached the mathematics advisers of three local education authorities and asked them to suggest secondary school teachers who, in their opinion, encouraged discussion in the classroom. The first research finding was the difficulty each adviser had in trying to think of more than one or two appropriate names! In the end we settled on intensive visits to 4 teachers in 4 very different schools: a small selective rural school, a large comprehensive school in a predominantly middle-class area, an inner-city school with a large majority of pupils from first generation immigrant families, and a comprehensive school serving a mixed catchment area of farming and mining communities.

In each case a first or second year class was chosen for observation, and one or more small groups of children within each class chosen as the site of our tape recorders. They were told that they were not being tested in any way, that the tapes would not be heard by their teacher, and that we were merely interested in how pupils talked about mathematics.

In each of the lessons we study, an observer writes notes, on a time sheet, of the activities of the pupils so that their actions can then be matched with the verbal recording. Later we interview the group of pupils to try to ascertain what understanding they have gained from the lessons.

Defining mathematical discussion

Like Derek Smith's group we found it easier to find definitions of what discussion is *not* than to arrive at our own definition of mathematical discussion. Like them we wished to exclude teacher-led talk which is merely a rhetorical adjunct to exposition, pupil-talk which is merely one person talking *at* another or to themselves, and talk which is about social ('I don't like this teacher') or administrative ('What have we got for homework tonight?') aspects of their mathematics lessons. At the meeting of the British Society for Research into Learning Mathematics in March 1986 we put forward the following which has remained our working definition ever since.

Mathematical discussion is:

○ *purposeful talk*
   ie there are well-defined goals even if not every participant is aware of them. These goals may have been set by the group or by the teacher but they are, implicitly or explicitly, accepted by the group as a whole.
○ *on a mathematical subject*
   ie either the goals themselves, or subsidiary goals which emerge during the course of the talking, are expressed in terms of mathematical content or process.
○ *in which there are genuine pupil contributions*
   ie input from at least some of the pupils which assists the talk or thinking to move forwards. We are attempting here to distinguish between the introduction of new elements to the discussion and mere passive response such as factual answers to teachers' questions.
○ *and interaction*
   ie indications that the movement within the talk has been picked up by other participants. This may be evidenced by changes of attitude within the group, by linguistic clues of mental acknowledgement, or by physical reactions which show that critical listening has taken place, but *not* by mere instrumental reaction to being told what to do by the teacher or by another pupil.

Although Derek Smith does not give a definition in such explicit form, the remarks made by his group show that they attach similar importance to the common purpose of the group and to their interaction. It is interesting to note that when instances of talk occur which fit this definition, they are often interspersed with non-mathematical social remarks ('Where are you going on holiday at Easter?', 'My guinea-pig has had babies', …) and we actually suspect that such talk might be a necessary element in enabling many pupils to sustain mathematical discussion. The classes observed so far, have provided sufficient material to allow us to categorise excerpts of mathematical discussion from three different points of view, and record these in the form of three parameters.

Classifying discussion

The first parameter deals with what it is that gives the speakers something to talk about. The parallel in Derek Smith's report is the statement that discussion demands *'relevant combined knowledge'* but our examples suggest that it is necessary to distinguish between different kinds of combined knowledge. We decided upon three categories:

a They have a task or concrete object as the focus of their talk;
b They do not have an understanding of something but they know this and it gives them something to talk about;
c They have some understanding and this gives them something to talk about.

The second parameter concerns the kind of language used in the discussion. Here we focus not upon the content of statements made but upon the language in which the discussion is conducted. Again there are three categories:

f They lack appropriate language, they do not have the right or useful words;
g They use ordinary language;
h They use mathematical language.

We predicted that sometimes the choice between g and h would be in doubt, since what is 'mathematical' language for young children might be 'ordinary' language a few years later: words like 'add', 'triangle', 'decimal' provide examples. Viewing the statement in the context in which the child was working, however, we find little difficulty over the decision.

We must emphasise here that it is the 'discussion' and not individual utterances which are being analysed. This, of course, is not always easy since pupils in a group may have different degrees of understanding and levels of language.

The third parameter derives from the kind of statements being made. This is the most subjective of our three points of view, and we feel it is likely to be closely bound up with the understanding being developed by the pupil.

p Incoherent statements; that is, incoherent to other pupils;

q Operational statements on what to do or how to do it;

r Reflective statements offering explanations or attempts to move beyond the immediate task.

How do we apply these categories?

Having done several examples using numerically specified lengths, the class has been asked to find the area of a trapezium with parallel sides, lengths j and k, and height h:

F j and k are....

R I can't do this...

G You can't do it....

R Do you have answers?

F You haven't got any numbers.

R How do you do it if you don't know what j and k is?

G Area....

F j minus k.

R How do you know what j is?

F No, I don't know.

G How do you know what j is then?

R How do you get j?

F And then you times it by h.

R You've got to have an answer.

G You can't get an answer.

The fact that the group effectively go round the discussion twice is good evidence that they are acutely aware of what it is that they do not understand. Their language is judged not to be mathematical because they are not using 'j' and 'k' as generalised numbers. Their statements are all concerned with how to do the problem and we would, therefore, classify the episode (b, g, q).

Transcripts are not enough

In most of our analysis, however, it is the participation as a 'fly on the wall' in the lessons which we feel to be crucial to our investigation of the relationship between discussion and understanding. The following extract illustrates this need to know the context of a particulr discussion. *These pupils are making -ominoes with square tiles. Earlier in the week they had been working with the investigation 'frogs'.*

A How many number of different moves did you get when you got to 5?

L You should have wrote yours down.

A I have....

A How many number of moves – how many number of things did you get?

C For 2 it's 1.

A How many did you get for 5?

C Just put 4, because that's about it, isn't it?

A It's about 5 isn't it?

L You can have loads of them.

This exchange is purposeful, mathematical and interactive. It is specifically concerned with the task in hand, but the striking feature is the inappropriate language being used. This investigation has nothing to do with 'moves'; the terminology is a hangover from the previous work. Nonetheless, the statements are quite coherent to the pupils and operational in nature, although the final remark by Len could lead them into a reflective phase. We would code this extract (a, f, q).

In the first extract in this article, not only was the presence of the observer in the classroom necessary to define the episode as discussion, but it also affected our classification. We observed that Tessa had been sent to Zara for help, and that the basis of the ensuing discussion was a pupil with understanding enabling another pupil to understand. We would code this extract (c, g, q).

Conclusion

No mention has been made in this article of the other side to our work: discerning and evaluating the pupils' mathematical understanding [1], but we wish to end with a note of caution, as much to ourselves as to others. If you are interested in the language and talk of pupils, *per se*, then the analysis of transcripts is an excellent way to do this. If, however, as we are, you are attempting to assess the concept building which accompanies the discussion, then it is important to realise that inadequate language and incoherent speech may mask real mathematical thinking and we should not make negative judgements about pupils based only on what they are able to articulate. ∎

Mathematics Education Research Centre,
University of Warwick

Reference
1 S E B Pirie, Understanding: Instrumental, Relational, Intuitive, Constructed, Formalised...? How can we know?, *for the learning of mathematics*, vol 8, no.3, 1988

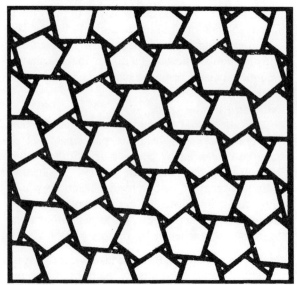

Chinese lattice derived from regular pentagons.

THE POWER OF STORY: ITS ROLE IN LEARNING MATHEMATICS

Rachel Griffiths and Margaret Clyne

$$Find \sum_{n=0}^{63} 2^n.$$

Once there was a king who was tired of war.

Story telling is a universal and age-old form of communication, which appeals to children and adults alike. Stories have been used since ancient times to explain natural phenomena, teach morality, raise issues and assist in solving problems. They provide a powerful means of communicating ideas and concepts, of raising issues and posing problems. Stories simplify experience, by focusing on particular aspects and by structuring the experience, and also enrich the experience by suggesting analogies and layers of meaning.

It is generally recognised that human relations, geography, morality and other aspects of life are embedded in stories, but mathematics has sometimes been regarded as outside the realm of story: a purely intellectual abstract discipline. On the contrary, mathematics, as an integral part of human experience, is also an integral part of story. Building on the mathematics which is implicit or explicit in a book or in an oral telling can assist children in developing concepts, solving problems and making connections.

Since 1985, we have been working with teachers and children in a large number of schools, using children's books to stimulate and develop mathematical thinking. We have found this approach to be extremely successful, providing both motivation and challenges for children.

Stories are valuable in mathematics learning in that they can:

○ *provide a context*
It is well established (eg Donaldson [1], Hughes [7]) that children can operate mathematically in meaningful contexts where they are unable to operate in the abstract, or in contexts which do not make sense to them. Stories provide familiar and meaningful contexts for children to explore, manipulate and make sense of mathematical ideas.

○ *provide a model*
Children can take the mathematical ideas from a story, and use the ideas in a story or situation of

their own. In this way, they explore the mathematics further, and assimilate the concepts.

○ *pose a problem*
Many stories pose a problem, either directly or implicitly. Through their involvement in the story, children are intrinsically interested in solving the problem.

○ *stimulate an investigation*
Stories can provide stimuli for investigations into different aspects of number, measurement, statistics, and applications in different areas of experience.

○ *illustrate a concept*
Stories can illustrate and illuminate mathematical concepts such as counting, measurements, permutations and combinations, logic and paradox.

Particular books or stories may do one, or usually more than one of these.

The power of story lies in the appeal which it makes to the emotions and the imagination. A story is more than a context, important as that is. Story involves a setting, conflict, tension or problem, and resolution of or solution to that conflict. In capitalising on the power of story to involve the imagination and to stimulate thinking, the mathematical problems or investigations posed need to be embedded in the conflict, tension or problem in the story.

Kieran Egan [2] argues for story as the medium for all teaching. He claims that stories are powerful in teaching because they depict binary opposites which are then resolved. In mathematics, he cites a story about counting an army of tens of thousands using only fifty pebbles. Such a story not only demonstrates the use of decimal place value notation, but also illustrates *the mathematical way of thinking: the triumph of ingenuity over cluelessness, of order over disorder'.*

The particular value of stories in mathematics is that they *tie the computational task to the human intentions, hopes, fears etc. that generated them in the first place.* By humanising, or rehumanising, mathematics, stories can return mathematics to its rightful place an integral part of human life.

Return to the two sentences at the start of this article:

$$\text{Find } \sum_{n=0}^{63} 2^n.$$

Once there was a king who was tired of war.

Which will motivate most readers to continue? The second sentence is the start of the story of King Kaid of India and the invention of chess. The king was so pleased with the game of chess that he offered the inventor any reward he wanted. The inventor asked for one grain of corn on the first square, double that on the second, double that again on the third and so on up to the sixty-fourth square. The problem inherent in the story is to ascertain the amount of corn the king will need to pay the reward. Thus, the story leads directly into an investigation of large numbers, of doubling and of the solution of the first sentence.

Year 6 (11 – 12 year olds) and Year 8 students have remained actively involved in finding the number of grains of corn on the chess board for over an hour (Griffiths and Clyne [63], Lovitt and Clarke [8]). In the process they have:

○ developed and tested hypotheses
○ found patterns
○ used a variety of recording methods
○ practised multiplication and addition
○ developed short cuts
○ gained deeper understandings of large numbers
○ investigated names for large numbers
○ discovered limitations of electronic calculators
○ used scientific calculators and explored scientific notation (Year 8), and
○ reflected on what they had learnt.

Year 6 children wrote the following comments about the activity:

We discovered if you multiplied the end number by itself [you]. get the number that goes in the row below. Then you multiply the answer by the top end one till you get the bottom number.

Hans says that it was fun because it was tricky and it made you think. I thought it was interesting because I learned a bit about chess. We also learned that the calculator only went to a certain number and then it stopped. We knew that because an 'E' showed up on the screen. When it did the 'E' I didn't know but I got most of them wrong.

We thought it was a fun activity because you really had to think about it.
We ran out of calculator space at the number 67, 108, 864.
We found out that in every column the last number was the same.

Steven thought it was okay. I thought it was okay too because we found out some things that we already knew like the calculator ending. We also found out that when we had worked out a few

numbers there was a pattern. The end number of each number was either 2, 4, 8, 6 in that pattern.

Further work on *King Kaid* can include estimation and calculation of the mass and volume of the corn, and their relation to the size of known large buildings, towns or countries.

Counting on Frank is a recent book which shows a boy engaged in mathematical exploration of his environment, as a reaction to the restricting nature of his home and family. Children and adults enjoy the humour of the book and recognise the pathos of the boy's situation, in which his dog Frank is his only companion. They are also intrigued by the calculations in the book, for instance that ten humpback whales will fit into his house, and only one-tenth of Dad into Mum's portable television. Asking children to check the calculations and estimates in the book results in many kinds of mathematical activity. Most readers are inclined to take the figures in the book on trust (after all, they are in print), and even to doubt their own calculations when these are in conflict with those in the book. As they work through the problems posed, they find themselves involved in

○ clarification of the problems
○ estimation of quantities and measures
○ measurement of length, area, volume, capacity
○ computation and calculation involving multiplication and addition, averages, large numbers, area and volume
○ ratio and proportion
○ making and checking hypotheses
○ explaining and justifying solutions
○ creating and solving new problems.

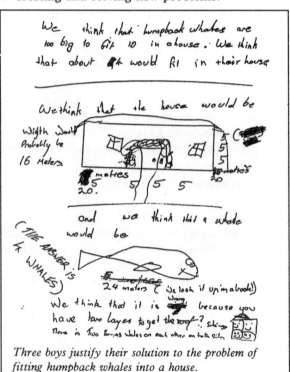

Three boys justify their solution to the problem of fitting humpback whales into a house.

Jodie's first attempt shows isolated points on her 'map'.

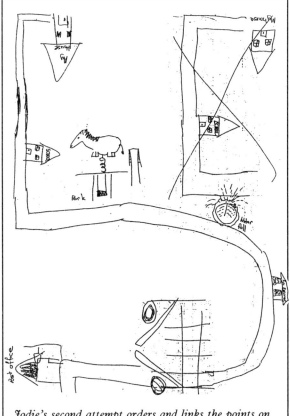

Jodie's second attempt orders and links the points on her 'map'.

The wide range of counting books now available provides a variety of experiences and activities, going well beyond counting from one to ten. Number is used in different ways – to label, to sequence or order, to count a set of items, to measure time, to count forwards or backwards, to show combinations and even to multiply. Similarly, the themes or narratives in counting books are diverse, and enable teachers to find books to fit topics such as animals, the zoo, Christmas, birthdays, conservation, satire, spiders or insects. The illustrations provide opportunities for classification, pattern work and paper engineering, as well as for exploring art techniques such as collage, water colour, cartoons and photography. Griffiths and Clyne [5] provide an extensive bibliography of counting books.

A counting book which uses ordinal, rather than cardinal, number is *Claude Money*, the story of the *sneakiest train robber ever*. Children can mime Claude's actions as he moves down the train, *behind the second carriage ...* They can build a train with cardboard boxes or blocks, labelling each carriage in numerals and words, and act out the story. Children can draw each carriage of the train, showing clearly its number and how Claude Money is moving in that section. They can match the text to each picture, and place the pictures in the correct sequence. Finally, they can write a new text modelled on this book.

Other books provide opportunities for mapping. *A lion in the night* shows a map of the round trip taken by the *lion who was stealing the baby*. After discussing the directional vocabulary in the text, children can go outside, make their own journey around the playground, and draw maps of their own journeys. In *The shopping basket*, Stephen passes a number of significant points on his way to the shop. Children can identify places of interest between their house and their local shop, and draw maps to show these and their route. The study of atlases and globes can follow sharing of books such as *A pet for Mrs Arbuckle*, in which Mrs Arbuckle travels the world in search of a suitable pet, only to find that the gingernut cat who accompanies her is ideal for the purpose.

Using stories in this way raises a number of issues and questions. Concern has been expressed that this is *disembowelling literature for the sake of mathematics*. It is true that teachers who attach to a story trivial or boring mathematical exercises will make children antagonistic to both the story and the mathematics. Any curriculum area can be trivialised. However, teachers who are sensitive both to the text and to the needs and interests of the children will assist children to reach a deeper understanding and enjoyment of both the text and the mathematical ideas inherent in the story.

A second issue concerns the movement from embedded to disembedded thinking (Donaldson [1]), from the context of the story to applications in other contexts and to generalisations. Related to this is the movement from informal language and recording to formal mathematical language and recording. This movement will not occur spontaneously. To assist this movement, children will need:

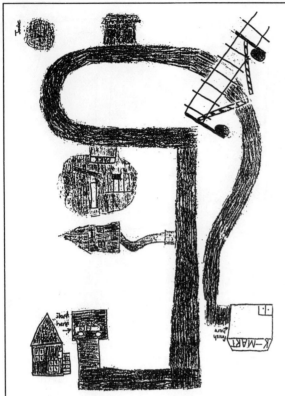

Jodie's final 'map' showing significant points on the way between home and the shop has been modified from her first attempt to incorporate conventional map features.

References

1 M Donaldson *Children's minds*, Glasgow: Collins 1978
2 K Egan *Teaching as storytelling*. Ontario The Althouse Press 1986
3 WP Galvin *The mathematics of 'The dam busters',* *School mathematics journal*. Newcastle: Newcastle Mathematical Association 1986
4 WP Galvin *Rosencrantz, Guildenstern and the Monte Carlo fallacy, School mathematics journal*. Newcastle: Newcastle Mathematical Association 1984
5 R Griffiths and M Clyne *More than just counting books: curriculum challenges for children.* Melbourne: Nelson 1990
6 R Griffiths and M Clyne *Books you can count on: linking mathematics and literature*. Melbourne: Nelson 1988
7 M Hughes *Children and number*. Oxford: Basil Blackwell 1986
8 C Lovitt and D Clarke *The mathematics curriculum and teaching program activity bank* Volume 2. Canberra: Curriculum Development Centre 1988

Stories referred to in this article

P Allen *A lion in the night*, Puffin 1989
M Anno, *Anno's hat tricks*, Bodley Head 1984
M Anno, *Anno's mysterious multiplying jar,* Bodley Head 1982
M Anno, *Anno's three little pigs*, Bodley Head 1985
P Brickhill, *The dam busters*, Evans Brothers 1951
J Burningham, *The shopping basket*, Collins 1983
R Clement, *Counting on Frank*, Collins 1990
A Graham, *Claude Money*, Era Publications 1987
G Smyth, *A pet for Mrs Arbuckle*, Ashtons 1982
Traditional. 'King Kaid of India' in *The Victorian fifth reader*, 1930 (facsimile reprint 1986)

This is an edited version of a paper presented at the 13th biennial conference of the Australian Association of Mathematics Teachers, Hobart, July 1990 and it also has been published in The Australian Mathematics Teacher Vol 47 No 1, 1991 – Eds

○ varied experiences of the concept or topic
○ practice with a variety of materials and methods
○ opportunities to share their ideas
○ demonstrations from other children and adults of ways of dealing with the concept or topic
○ challenges to their thinking, to stimulate thought
○ opportunities to reflect on and refine their ideas.

Another comment is that, whilst this approach works well for early primary students, it is surely not appropriate at higher levels. However, we have ample evidence from our experience that students in upper primary and lower secondary years are highly motivated by stories such as *King Kaid of India, Anno's mysterious multiplying jar,* and *A pet for Mrs. Arbuckle,* to explore mathematical ideas. *Anno's three little pigs* and *Anno's hat tricks* are two picture-story books with great potential for the development of mathematical thinking for older students.

Galvin [3], [4] has given other powerful examples, such as *The dam busters,* to use in upper secondary classes. The popularity of authors such as Martin Gardner, Raymond Smullyan and Douglas Hofstader is testimony to the appeal of story in illuminating and stimulating mathematical thinking at all levels. ■

Ministry of Education, Victoria, Australia

TOWARDS A LANGUAGE
OF STRUGGLE

Mathematics and language are closely related forms of life. They are both symbol-systems. They are both action-systems. Both use classification and transformation as organising principles. But language is acquired by almost all very young children without the help of teachers. While mathematics is acquired by most children very slowly, in a contrived way, and only a little mathematics at that. If these are two closely allied forms of life, then why is one acquired so naturally, the other not?

The most obvious difference between the place of mathematics and that of language in our lives is that language has become vital for the daily survival of the species, in a way that mathematics is not and could never be. Thus, for example, babies hear adults and older children using words every day, but not speaking in a specifically technical mathematical way. Almost every child has an overwhelming need to learn to talk, but far fewer need to learn formal mathematics. Society is so arranged that opportunity often matches need: most babies learn to speak at home; at school the learner has to share his teacher with about thirty others.

Granted these most obvious differences, why is it that when they do come to school, so many children find it so hard to build a mastery of mathematical forms, using those same mental powers that they have already displayed in acquiring spoken language?

Two reasons why mathematising becomes inaccessible to a large number of children in school seem to be firstly that mathematical action is in itself relatively difficult to take part in; and, secondly, that children are frequently presented with tasks in the name of 'doing maths' which are trivial, worrying, or just incomprehensible.

Spoken language has many similarities with mathematical forms of expression. This suggests the argument that anyone who has learned to talk ought to be well placed to mathematise. It is worth looking more closely at three related aspects of mathematical expressions and of everyday language:
tolerance, ambiguity, clarity.

Everyday speech is a highly tolerant medium. This tolerance is necessary because conversation is a form of action in the world; it permits people with differing perceptions to get on with the action, as long as they take anothers' main drift. It allows children to join in, and to learn how to talk by talking. Because it is a tolerant medium, everyday language is necessarily ambiguous. It is only when one looks closely at the meanings of words - in philosophical arguments, for example - that the simplest remark turns out to be full of holes. Everyday conversation is easy to understand, its meanings are clear, because we speak in the context of everyday. When we are in a bus shelter, and a bus comes round the corner, the words 'Here it is' have a clear meaning.

Without having to think about it, we frequently exploit the clarity of everyday conversation by making the simplest remark carry many layers of less obvious meaning, like a kind of halo. In this sense, everyday speech reduces communication to a maximum.

Now, mathematising is clearly also a form of action in the world. And its expressions, however carefully defined, have to retain a fundamental tolerance. As a result, people with differing perceptions of them can agree on shared uses for them. And children can join in mathematical activities before they have grasped all their aspects, connections and consequences; and learn to mathematise by doing it.

Because it is a tolerant medium, mathematics is also necessarily an inherently ambiguous one. It seems that many children find the ambiguities of the mathematical expressions they meet too much to cope with; especially perhaps because these expressions so often have other expressions as their immediate context; whereas everyday language has the everyday world as context.

Divorced from the world, mathematics is free to become a more self-controlled and self-centred form of action than everyday conversation. It acts on itself, while everyday talk acts in, and on, the world. It is characteristic of mathematical action that its symbols are pressed harder than we commonly press words in the casual process of the day's talk. Mathematising involves an awareness of one's moves, whereas in talking words are commonly used unselfconsciously.

Because mathematics provides its own context, and because the structures involved soon become too complex to hold in the mind, mathematising can only develop beyond a simple level if it is mediated by signs. Speech is a primary form of expression, of which writing is a secondary, recorded form. In mathematics it is often the case that non-verbal symbols are the primary form of expression, and that any talk, written down or not, which accompanies this notation is secondary.

In so far as mathematics is founded on awarenesses which are essentially spatial, pictures, diagrams and models provide us with more direct and powerful expressions of these awarenesses than words can.

$(a+b)^2 = a + 2ab + b^2$ What is the relationship of this sentence to the awareness of which it is an expression?
Consider the following steps:

Arithmetic (numerals structured by an algebra);

underpinned by

Algebra verbal, linear, temporal & sequential;

underpinned by

Geometry—pictorial, non-linear, spatial & total.

When they first come to school, children are presented with mathematics in a verbal form: 'This shape is called a .... ?' 'Find the set of red things'; '3 + 7 = ?'. These forms do not connect in any obvious way with a child's image-making power or suggest the need for a spatial form of awareness to him. And yet it may be that, in order to cope confidently with numbers, shapes and so on, the child needs to make such connections. (Compare the 'mind-pictures' proposed by Mary Boole.)

Therefore, at the very time when young children are being invited to struggle with the relationship of speech to its printed signs, they are also asked to cope with the very different relationship of mathematical awarenesses to printed words and to other non-verbal signs. In the sentence '2 + 3 = ... ?' for example, the signs all look deceptively like one another, and like letters in a printed word: what grounds for confusion are there here?

There is an aspect of doing mathematics which is more like writing a poem than it is like talking. When trying to write a poem, one struggles with the possibilities and consequences of particular phrases; when trying to work at a piece of mathematics, one struggles with the possibilities and consequences of choosing certain signs, or a sign-system; in both cases, one is trying to capture awarenesses.

The expressions of mathematics, like those of the poet, tend to be more exacting than those of everyday speech, for in both cases one has to struggle with the difficulties of one's medium, in order to achieve new clarities. The hoped-for outcome is a hard-won and particular realisation of awareness, expressed symbolically through the achievement of a fresh relationship within the sign systems used.

In mathematics, such processes require a language of struggle. The history of mathematics is full of examples of creative mathematicians struggling to hammer out agreed usages, in the course of working on unsolved problems. Expressions which achieve a clear sense for a given group at a particular time, emerge out of the language of struggle. An agreed solution to a problem creates a space within which it is possible to operate. Ambiguities are pushed aside, at least for a time, and for that group which participates in the agreement. It is the common fate of such agreements eventually to be undermined, abandoned or supplanted, in the process of arriving at fresh solutions to problems.

There is a form of mathematical presentation, which is a by-product of the creative or problem-solving process. When tidying up their work, mathematicians may invent some clearly-defined terms and axioms, and proceed to demonstrate what has to follow, given such starting points. Such formalised demonstrations can only be really useful to someone who has access to the mathematical actions to which they refer, though at a sophisticated level the very form may assist access.

Unfortunately, text books and teachers frequently use tight and formal language when they present mathematics to children. Such a presentation appears to be an echo of the formal demonstration. It gives no hint of the language of struggle through which mathematics is created and owned, whether by adults or children. Thus, the seemingly precise and clear words of the text book may—in the event of someone understanding them—nevertheless, relate in an obscure and muddled way to the mathematical actions to which they refer.

The cure for this disease of text books cannot be to insist on ever increasing precision in the defining of terms, since such pedantries have little or nothing to offer the child who does not understand what he is being invited to do. It should be clear that insisting on the so-called correct language from the start is a dangerous exhortating. 'Line segment' does not clarify for a young child, whereas 'part of a line' involves images on which he can work. The problem of how to make mathematical actions more accessible to more children is a profound and complex one, which can only be tackled at all by recognising the need for the language of struggle, out of which a different kind of clarity is born.

However sensitively composed, the language and symbols of a text book or work-card can only be an invitation to mathematical action; clearly they cannot guarantee that it will happen. When attempting to work at mathematics with children in school, therefore, talk is also essential. Any group of learners has to struggle to achieve with their teacher some shared mathematical forms of action, to which both the learners and the teacher have genuine access. This is another aspect of the language of struggle, involvement in which permits the teacher to begin to learn about the learner's mathematical awarenesses. For example, when talking about a mathematical

problem with a child, informal everyday language can easily add its own ambivalences to those of the system being used. However, since this language is now pointed towards the mathematics and not towards the everyday world, it too may begin to behave in a special way in this special context. For the teacher has to listen, watch, and then to respond with more than everyday sensitivity if he is to begin to grasp what the child is showing that he knows. In other words, it is important for the teacher to respect what the child can say and do, rather than assuming that these are necessarily to be improved. For it is often more appropriate to help the child to develop his own competence, rather than to invite him to begin at some point within the teacher's competence. And it can be fruitful to encourage a child to pursue his own speculations when conversing with him, rather than muscling in with knowledge of one's own. In this connection, it is worth asking how often a teacher's 'explanation' is a truly helpful response to a child who appears puzzled or confused. Paying attention to another person is hard work. Especially in a crowded classroom.

COMMENTS ON LANGUAGE
IN THE CLASSROOM

It may seem easy to talk about language as an entity separable from the natural context of the events within which it takes place and of which it is also an aspect. But it is dangerous to do so if we wish as teachers to gain insight into ways of knowing about language in the classroom. The pressure in a situation is the need to evoke action in response to the felt demands of the moment. Effective action depends on an immediacy and richness of response that may actually be hindered by the belief that there is something called 'language' which can be separated from events.

An experience commonly a surprise to teachers is that of finding that a topic which went well with one class unaccountably fails to take off with a different group of children. This is a reminder of the limited value of a data transmission model of communicating in a classroom. One danger of attempting to take action from such a model is that it leads naturally to over-stressing the value of clear exposition for the development of a worthwhile classroom situation.

An account of the language used in a mathematics classroom will leave out much of what made the situation viable and the learning effective. Such an account may focus on the clarity of the exposition which enabled the pupils to grasp what was being taught. But this may be of no help to a skilled teacher who attends to techniques for achieving clarity yet still finds difficulties. The account may also carry with it, by its very existence as an account of that kind, the implication that the clarity may be available to the same teacher with the next class which comes to his room. This would be a mistake, one which is rooted in the complexity of language and activity. There are implications in these comments on language for the possible value of theoretical studies of language for the teacher. The value is likely to be indirect, rather than something that provides a direct methodology to underpin successful teaching, but theoretical studies can have a tangential effect, helping the teacher to develop richer ways of responding in the classroom.

My own memory as a pupil of irritation with a teacher who spoke too much and interfered by doing so with my attempts to come to grips with something new reminds me that in the complexity of a situation in which learning takes place, an attempt by the teacher to be clear, definite and unambiguous may cause more damage than remaining silent.

No retrospective criticism of that teacher is intended. He may have been aware of the limitations of his attempt to speak to 30 children at once. What he said may have been as much a wish to encourage his class to work at whatever he had brought to them. The change from taking a class as a single entity to working with smaller groups facilitated by, for example, grouping the desks together is a different attempt to make available an encounter with the children. The natural language of such an encounter is of a more conversational kind than that of the public speech necessarily made to no one in particular.

The phrase 'school mathematics' is often heard. It might be comforting to believe that it was always used in the same way by teachers; but it can be revealing to encounter the variety of interpretation by teachers. It would not be unusual to find that two schools, whose only mathematics textbooks are, say, those of the School Mathematics Project (SMP), will produce pupils who are quite unlike in their mathematics work. One might, as an observer, notice that in one school the text is made the basis of exposition by the teacher and a natural consequence is that the children do the exercises. It is likely that this would be a school that would be quick to buy the SMP supplementary exercises which have recently become available. Sections in the books devoted to 'investigation' will be ignored, except perhaps occasionally, 'for a change', or 'to help get over the concept of...'. In contrast at another school the textbook will be more a resource book for the teacher than text for the pupil. The experiment sections may be cut out from the book and stuck onto card to be available as individual tasks to be worked at by just a few children. This may be dismissed by some as no more than a change of style, and this will be the case where the way in which a particular teacher works embodies his wish to be the sole authority in the classroom.

It may be noticed in this case that there will be little attempt to promote tasks with a variety of interpretations and consequent outcomes, neither will children be encouraged to help one another or to challenge each other with tasks of their own —in short to promote a growing independence of the teacher.

It could be, however, that the different use of the textbook is part of the setting up of a different structure for a wider range of learning and doing for the children, as well as for the teacher. The place of concrete material could be different. No longer just 'visual aids' or pieces of structural apparatus to illustrate a concept, materials can be part of the starting point for many kinds of work.

There is more significance in sticking the contents of a textbook onto card than mere economy. If only a few children are working at a particular task, the pressure towards reaching a common interpretation and a single outcome is avoided. The confidence of children to take action in their own ways is given support by the way in which work is made available. Where the teacher intends this kind of growth in the children there is enormous scope for enhancing children's authority with respect to what they do. This is in contrast with a language of exposition, where the outcome is doing exercises. There is then quite a different relationship between The language and the consequent action of the child.

Some descriptions of how language works may fail to make the distinctions between different forms of language which have been hinted at. It might be said that since mathematics has to do with a self-conscious reflection on certain aspects of experience and activity, all mathematical work would have the same character. This view would neglect the place in classroom activity of the intentions of the child when he is said to be learning. The notion of intention seems to be an important one, not only as a significant personal characteristic to be studied but also as something to be provided for within a classroom. The notion is too often reduced to 'the problem of motivation'. This fits in with a view of the classroom based on exposition, but does not help when the wish is to ask how learning may be made more effective.

GENERATING PROBLEMS FROM ALMOST ANYTHING: PART ONE

Marion Walter

The theme of this conference is *economy* − so I did not bring any slides. So many of you are wearing clothes with wonderful geometric patterns − would you look around you and make up several problems? I am not sure why I prepared a talk − we could make up problems even just from Heather's three-dimensional earrings made from hexagons. I'd like you to move so that everyone sits next to someone who heard the talk I gave two years ago [26]. This activity made me think of a related mathematical problem − one I have not seen before. Suppose that there are 5 people a, b, c, d, e sitting in a row and you ask them to move so that everyone of them has at least one new neighbour.

What is the *least* number of interchanges that have to be made to accomplish this?

How will your strategy and answer differ if the 5 people sit in a circle?

What if there were 3, 4 or 6 people?

Well, this is an example of a problem suggested by an activity and of course you can change it many ways.

There are quite a few ways of coping with *economy* if we had no texts or materials. Here is a *partial* list of things (in no particular order) that we might consider. *Scrap materials, situations, an activity, given or collected data, drawings, surroundings* (geometric and consumer related) *equations, tape* (two kinds), *a theorem, a condition, an exercise, an answer, a proof, a counterexample, a problem, a game, a mistake.*

How will we do this?

Be patient − but we will only have time to look at some of these categories.

Scrap Material

One of the things I was going to examine today was a hundreds chart or grid because it is endlessly rich but since I found out here that ATM has put out a nice book [10] I won't. Instead, look at this paper which came from a chocolate box and you can see a different sort of grid on it and there are lots of problems there.

For example, the 4 rows on this paper have 3, 4, 3, 4 cups in them but the diagonals have 2, 4, 4, 3, 1 respectively. Suppose there were 5 rows or 6, then how many would there be in each diagonal?

Suppose the numbers from 1 to 100 were written in such an array. In which row would 79 occur? There are many ways you can use such scrap material.

A boy named Joel used square grid chocolate paper to make a lovely man and he wrote, 'My head is square made of 4 cm by 4 cm paper. My eyes are 1 by 1 (cm) My neck is small − hardly there at all. My body is plump and measures 10 cm by 10 cm ... Do you think I look like you?'

But first let us remind ourselves that we have no stuff, we have to make do with what we have. Suppose that we do happen to have only one lonely square. Well we can cut it in half and get two isosceles right triangles.

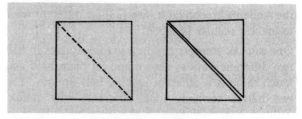

We can cut one of the triangles in half to get two smaller isosceles right triangles and we can do it again and again.

How many times can we do it before we can't pick up the pieces any more?

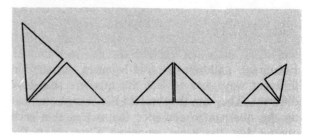

In theory everyone here can have a right isosceles triangle – and they are all similar to each other! What other 2-D shape(s), if any, can you find that you can cut in half to obtain two figures each similar to the original one? Can you find a 3-D shape that can be cut this way? It is not good having such triangles if we don't have a problem! If we put the two congruent triangles on top of each other; one High School-level problem is to find the area of the overlap [22].

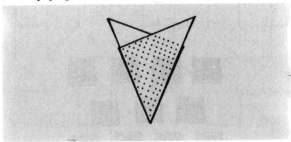

Try telling someone on the phone how to place the triangles as they are shown in the diagram! Young children can make patterns and designs with the various sized triangles and even work with some special fractions.

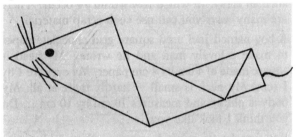

Suppose you don't even have a square; you have to make your own. How many different ways can you make a square? Draw one. Construct one with straight edge and compass.

So many books show only one way and then they tell the students how to do it. See [23] for different ways to make a regular hexagon.

One can ask younger children to fold squares in half two different ways as shown and let them hold and touch these two different looking half squares. The two halves certainly look different; why should children believe right away that they have the same area?

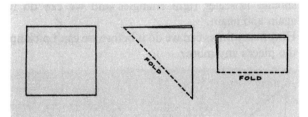

Encourage children to fold quarter squares in different ways and to feel the quarters that look different. They can then unfold the squares and cut up the quarters to convince themselves that each quarter has the same amount. What about the rectangular paper mats we have at lunch and that get thrown away after each meal? Try folding the diagonal(s) to see halves and quarters. How should a young child know that the 4 parts are equal in area? It is a nice exercise to show, by cutting and rearranging, that the 4 triangles do indeed have equal areas.

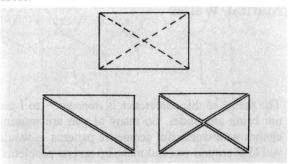

One problem that suggests itself when one folds a diagonal for the rectangle and then folds the flaps that cry out to be folded is to calculate the fraction of the rectangle that a triangular flap represents. Now fold the rectangle again so that the diagonal that was creased before; folds onto itself. The flaps again turn under which comes as a bit of a surprise though it is 'obvious'.

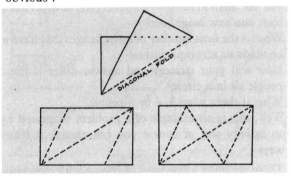

What is the shape in the middle – are you sure? Can you express the area of this rhombus in terms of the length l and width w of the rectangle? [24]

Doing problems in more than one way

If only a few problems are passed on in an oral tradition when all textbooks disappeared, which ones would you 'save'? Well you might want to collect ones that can be solved in several different ways. Which is your favourite? One of my favourite ones is the well known handshake problem:

If 5 people all shake hands with each other, how many different hand shakes are there? *(See Jeannie Billington's and Pat Evans' article on p12 of this issue for a discussion of the handshake problem – Eds.)*

Another problem, which has appeared in many different places, is to find the area of the petals in the diagram:

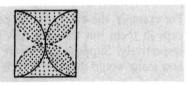

Using the circular paper coasters we have at lunch, we can cut semicircles and lay them out in a square to see the overlap on the overhead projector. It is a problem that can be solved in many different ways and changed in many ways.

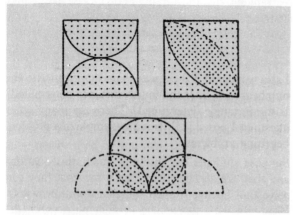

How else can the problem be changed? There are many ways — and we will see some methods later. For now, note that we began with a square on the outside and 4 semi-circles inside. You get a different picture if you put a circle on the outside and 4 half squares on the inside.

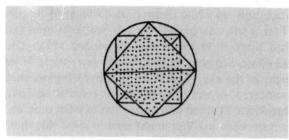

Are there other ways of doing this? We can pose the problem; what is the area of overlap? Can you solve it in several different ways? But can we also think of some new questions to pose? Note that here we turned the diagram inside out in a sense — and one way of changing a problem is to interchange parts!

For younger children it is a problem just to analyze and copy such designs. Looking at these pictures makes me realize that we can get lots of problems from pictures.

Problems from pictures

One way to economise is to give children not a problem, but just the diagrams and have them make up their own problems from the diagrams. For example, if the picture is

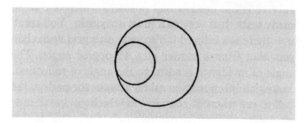

what problems might they or we pose? The picture here is simple, it could not be much duller but we can still pose some problems. Can we suggest some now to get started?

○ Which is the bigger circle?
○ How much do you have to roll the inner one to get back to the starting place?
○ How many small circles fit into the large one?
○ What is the area of the crescent?
○ What is the circumference of each circle?
○ What is the ratio of the area of the small one to the area of the large one?
○ How big does the small circle have to be so that the ratio of their areas is say 1:3?
○ If you draw an arrow on the tangent of the small one before you start rolling it where will it point as you get back to the starting point?
○ What if the picture were of a three dimensional object?

Can you see how narrow some of the textbook questions are that ask just one thing? Then there is usually little thinking — students check the answers in the back of the book and that is the end of it.

If we put another picture beside the first one which is related to it — same size circles just in a different relationship, you will think of different problems. Even the same ones can excite new interest.

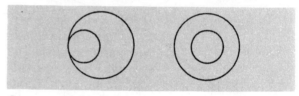

Can you think of some?
○ What if it is a picture of a three dimensional object?
○ In how many different ways can you place two circles?
○ Investigate the picture as the circle moves to the right and as one participant said — do you want your egg fried or scrambled?

Here is one of my favourite bad examples that goes with this kind of picture.

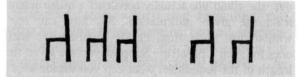

If 200 years from now the textbooks are found, people will think chairs were always arranged behind each other. What is *the* question that goes with this picture in most elementary textbooks? 3 + 2 = ? What are some other questions? Can you make up some?

A very large number of problems were produced by people at Marion's lecture. Some of these will appear in MT121. Let us know your questions — Eds.

Problems from making up problems

If one makes up problems or we ask our students to make up problems, we will find that we are often faced with new problems. For example, suppose that I am making up a simple geometry problem in which I want to have a parallelogram ABCD which is not a rectangle, where BE is perpendicular to AC.

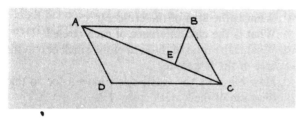

Now suppose we want to give as few angles as possible in order for students to find all the angles. How many angles do we have to provide? Which angles can they be? Is there any size of angle we cannot choose in order not to contradict the information? Might it not be useful for students to make up such a problem? I often ask my college students to make up problems for their class mates.

There is a problem which can be found in several textbooks. It is one in which one must be careful in choosing numbers. Here is an example of this type of problem.

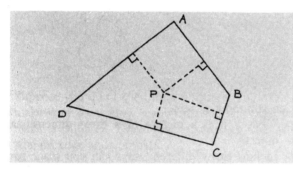

Line segments are drawn from a point inside a quadrilateral perpendicular to the four sides. The length of these segments and the length of the sides are given and the problem is to find the area of the quadrilateral. It is alas, easy to choose numbers that are impossible — as one of our good students found out, she could not actually construct a quadrilateral from the given information. So, how can one, without just measuring, decide numbers are legal ones for such a problem? Suppose you are given the length of the four sides, what can you decide about the length of the four segments for a given region for point P? I have not worked on the problem.

One problem I have worked on, which has appeared in various forms, is as follows: There is a rectangle ABCD with a point P inside the rectangle. The distances PA, PB, PC are given as shown and we are asked to find PD. APC is not a straight line. It is a nice problem that can be done in several different ways. When I first saw it, I wondered how one could find PD without knowing the length of the sides.

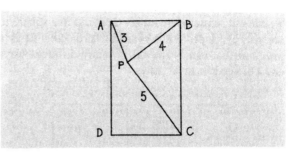

I also wondered if there was some significance to the numbers 3, 4, 5. Is the fourth distance determined? Is the rectangle determined? There are many other questions I posed [25]. It is an example of a problem begetting problems.

Suppose students are studying quadratic equations and they are asked to make up a few that they can solve (not having studied imaginary numbers yet). How will they choose the coefficients so that the roots they get are real? What will they have to explore — find out?

Problems from situations

We already saw that we can be led to a problem from the reseating we did earlier. Here is another problem that came up while flying from London to Seattle. This is what happened. We left Seattle at 9pm and had flown 3 hours so my unchanged watch said 12.00am. Someone looked out and saw quite a bit of light in the sky. Was it the Northern Lights, they wondered, or was it the moon, or could it be dawn breaking? That led to the question of what time was it now below us? You might need to know that there is an 8 hour time difference between Seattle and London and that it takes 9 hours to fly there. We might want to simplify the problem and assume that London and Seattle are on the same latitude and that we flew along a latitude rather than the polar route. What shall we do about the time zones? All kinds of questions arose among the passengers including ones about whether the earth was turning *and* the plane with it or the earth only but not the plane!

Making use of some problems that survived

Let's assume again that all texts, Xerox machines and journals etc. have disappeared and that the only problems that we know are the ones handed down from grandmothers to daughters and sons through the oral tradition. Which problems do you think would remain? One that I would like to think would survive is the billiard ball problem which is now in many texts. It is very rich in its own right. You recall it — there is a billiard table made on a grid and a ball gets shot from a corner at a 45 degree angle. The angle of incidence is equal to the angle of reflection. Young children may be asked to trace the path of the ball to see where it goes and in which pocket it ends up. Older children are usually asked to predict what pocket it ends up in for various size tables.

What are some other questions?
- ○ Does the ball always land in a pocket?
- ○ How many rebounds are there?
- ○ How many squares does the ball pass through?
- ○ How long is the path for a particular table?
- ○ Can you predict the length of any path?

You are too good! Most *texts* just ask in which pocket does it land? Realizing that two attributes of the billiard table are that it is rectangular and that the grid is square, you may want to ask 'What-if-not?'[4] Well, the grid may be triangular and you may also want the table to be triangular or you may want to try to keep the table as nearly rectangular as possible by making it a parallelogram. The same questions as above can be used or you may think of new ones.

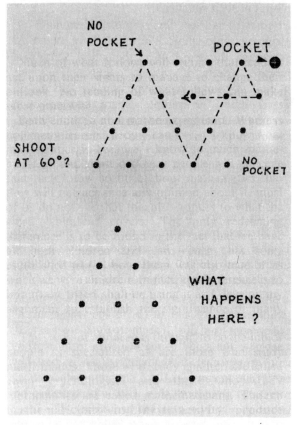

If you decide to change the number or locations of the pockets in the original or new tables more problems are available. These grid problems remind me of the rectangle problem I first came across in Starting Points. [1]

Now look at some of your favourite surviving problems and see how you can extend and change them! ■

(This is Part One of an edited text of Marion's exciting talk at this year's Easter Course at Nene College. The second part and bibliography will appear in MT121. If Marion spurred you on to tackle some problems, we'd like to hear from you — Eds.)

GENERATING PROBLEMS FROM ALMOST ANYTHING: PART 2

Marion Walter

In this issue Marion concludes her lecture given this year at the ATM Easter Course in Northampton. Part 1 appeared in MT120.

Maths from visualising

If there is great economy we won't have many pictures except the ones in our head (and new ones we paint). The *ATM, Leapfrog* [12] and *Dime* [9] material has been particularly rich in visualising work – see for example the *ATM* book *Geometric images* [27]. One of my favourite pieces in that book is the problem given to children: visualise 5 lines and a point and write about your picture. The children's descriptions are wonderful. Then there is the *Leapfrog* tape called *Imaginings*. That is one type of tape we can work with or get ideas from. (The other type of tape is 'ordinary' type tape or the kind of 'tape' with holes from computer paper that someone told me they found particularly rich). Let's do a bit of work with the ideas from *Imaginings*[12]. It is some time since I listened to it so – –

Close your eyes and visualise a line in front of you; a horizontal line and let a ball slowly roll along it. I don't think the tape says that you should let it drop off but you might as well let it drop off as it will bounce right back again. Now would you fold up one section of your line so that when the ball rolls toward that end it can't fall off. Now roll the ball in the opposite direction and fold up that end of the line so that it can't roll off. How does your picture look now?. If your picture looks a bit like an open box would you raise your hand? (Almost everyone does) If your picture made a triangle would you raise your hand (about 4 or 5 people did).

That raises a lovely problem. If you have a line, and let's simplify the problem and say you can bend it only at unit intervals, which bends will give you triangles? How many different triangles can you form this way? (Later you might want to ask what the probability is of obtaining a triangle if you make two bends at random). Let's take an example.

Suppose you have a 10 cm stick which can bend only at cm intervals, how many different triangles can you obtain? Young children can solve the problem by using straws that they mark at cm intervals. Even older students sometimes are powerfully reminded of the triangle inequality which they sometimes first ignore in counting the 'triangles'. It is perhaps surprising that one can get only two triangles and that they are both isosceles. If you work with different lengths to obtain data – it is not very easy to see what is going on! If enough data is generated older students may be able to notice that the number of triangles on length n (n odd) is the same as the number of triangles for length n − 3; that is $T(n) = T(n − 3)$ for n odd and where $T(n)$ stands for the number of non-congruent triangles that one can form with integral sides if a length of n is provided. The general formula is advanced math [11] and [18].

Let's look at one more of the visual exercises on the *Imaginings* tape and see how we can use it as a starting point for more problems. *Close your eyes* again and visualise a triangle – vertically in front of you. You might want to make it acute or obtuse – or skinny – though skinny is not a mathematical word. Eventually make it equilateral. Imagine it white. Now imagine a small equilateral triangle in each corner – imagine these black. The tape then asks what shape is the white piece? Could you let your small triangles grow and grow so that the resulting figure in the middle is a regular one – what regular figure does it become first? Can you make the black triangles grow a bit more so that the figure in the middle becomes a regular something else? What does it become? And if you grow the triangles more what happens then. Can we now use this visualising problem to make even more problems? Let's list some attributes of some of the pictures we saw. How could we describe it?

○ it has symmetry
○ it is a triangle
○ it has an irregular hexagon inside
○ the corner shapes are triangles
○ the corner shapes are the same shape as the outside shape
○ it is two dimensional

It is interesting to ask what if it were three dimensional? We would have a large regular tetrahedron with a small regular tetrahedron at each corner. If we cut off each of the small tetrahedra,

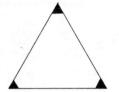   

what would the remaining shape look like? What if we first grew the small tetrahedra until they touched and then cut them off? What would the shape inside look like? Can you picture it? One way of generating problems is to list some of the attributes of your starting point – we listed only a few – and to pick one or two of them and ask, 'What if it were not so?' For example, what if it were not two dimensional, what if it were not a triangle? One alternative is that it is a square and that there are still triangles in the corner – not equilateral but perhaps right isosceles. Or there could be squares in the corner. We could ask the same questions as we asked before and we could again try visualising the shape in the middle and see what happens as the corner shapes grow. Of course you will also think of many other new questions suggested by the pictures. [4]

Scrap and other available material – again

When I tried to think of what material is either often thrown away or is easily available, several things came to mind. First, we always have some water available. What kind of problems can we have with water? Well, there are the well known Piaget type problems and experiments: which container holds the most, the tall skinny one or the short fat one? But there are also the supermarket deception kind of problems. One nice simple exercise you can do on your way out from this lecture hall. Look at the different bottles we have on the table and order them according to size just by looking at them.

I know that I have been fooled in a supermarket by 12 oz jam jars that looked like one pound ones and 13 oz 'pound' coffee tins. These containers are very cleverly designed to try and fool us. You might want to work on and explore these deceptive packages that look as if they hold more than they actually do and design packages that deceive so that you and your students become more aware of the deceptions practised.

Which holds the most, the tall skinny one or the short fat one?

Then there are all those water pouring problems that I recall I could do when I was 8 and that I now find more difficult! Given certain sized jugs, one is asked to pour out different amounts. Do you remember them? Perhaps you are given a 10 pint jar filled with water and an empty 7 pint jar and 4 pint jar, can you measure out 2 pints? So here is an example again of problems begetting problems because a first question is, is it possible to do this? If so, is it trivial? If not, why not?

What sets of three numbers make a possible problem? A good problem? What are the fewest number of pourings needed? I did not remember from my childhood that a container was filled with water. So, another question is, what if you were not given a ten pint container *filled* with water? Or what if you were given ten pints of water but not in a ten pint container? Or if you were given only the 7 pint and 4 pint container and a tap? Are all these problems different? Perhaps a group could look at pouring problems and how to make them up. And what about if you were given three empty jars? The only reference I could find was in Coxeter's book [7] and there trilinear co-ordinates and reflections in an equilateral grid were used to find a solution. That raises a question: is there any connection between the billiard ball problem and two containers?

Then there are the other old mixing problems – the kind where you have a glass of coke and some rum and you pour a cup of the rum into coke and then take a cup of the mixture and put it back into the rum glass. Is there now more rum in the coke glass or coke in the rum glass? And what if you don't stir well. And suppose you also have a jug of cream and use the three things to pour and mix?

Another suggestion – first made to me by Pearla Nesher – is to use food colouring and water to explore problems in ratio and proportion. If you use three drops of food colouring in 1/2 glass of water, how many drops should you use in 3 glasses to get the same colour? I am sure you will have many other water ideas – some dealing with ice!

Then you can make use of newspapers and magazines. You can use the graphs in the newspapers for data and explorations. You can, of course, use the papers to make rectangles and squares and circles. Use them also to make tubes which can be used for making structures – an old ATM activity. You can collect pictures that you can use for counting activities. You can cut out pictures of circular objects like cups that look like circles but when you cut them out and look at them they turn out to be ellipses. Trade marks are also a very rich resource. There is much work you can do with ads — look at the bad logic that is often used.

There are many kinds of cardboard tubes — kitchen paper towel ones, toilet paper rolls and xerox paper rolls which are very sturdy indeed. What can you do with paper rolls? You can print circles with them — useful for number chart work — circle the multiples of 3 etc. A sophisticated problem is to examine how they are made. Young children could just unroll them. During the first workshop I did with materials in Lancaster in 1971 we made plaster casts from them. You get the helix mark on the cast which is nice. You can also cut the rolls and rejoin them. If you don't cut perpendicular to the roll, the cross section is not a circle and you can join the two pieces in only two ways. You can make some nice shapes this way — and raise some nice problems — can you get the construction to close again for example?

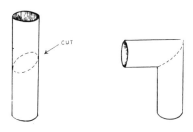

David Cain collected lots of identical cardboard tubes. Imagine that you will look through one and count how many horizontal rows of squares you can see on the screen. Can you predict how many it will be? Now look and see! Starting from the top row at the back of the lecture hall and working down row by row to the front; let's see how many you can see. 9,10,9,8,8,7,6,6,5,4,3,1. I think there is a problem or two here we could work on! Is there anything you could do — other than getting out of your seat that would change the number of rows you can see? Fewer? More?

What else do we throw away all the time? Milk cartons! There are many things you can do with them. Let's only look at the cartons that we have cut down to form cubes without a top. How many squares are they made of? One of the things you can ask is, 'How many ways are there of arranging 5 squares?' Here are just a few:

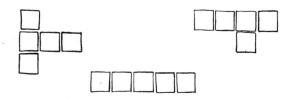

Some of them fold into open cubes and some don't. How many different patterns are there and how many of them fold into boxes without tops? Which patterns fold? Number the ones that do 1,2,3,... Now write one of these numbers on the bottom of your milk carton and try to cut it so as to obtain the pattern whose number you wrote on your carton. [19], [20], [21]

Here is another box with all sides congruent trapezoids.

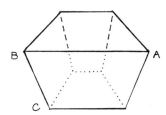

It raises all kinds of questions about the size of possible angles. What is the smallest that angle ABC can be? the largest? Is there a smallest? A largest? What happens in the extreme case? What is the volume? How much paper is used? Is it more or less economical to make than a cube?

Then there is much work to be done with cereal type boxes. Collect some of them and examine them for volume and surface area. You will be surprised! Also you will find some pairs of containers where the one with larger volume actually contains fewer cornflakes! Why do manufacturers make narrow containers that waste material is a question worth discussing with your students. You can also estimate the amount of material saved if say 10,000 boxes of a given volume were made with better dimensions. There is a lot of ordinary curriculum maths in this type of investigation.

There are different problems that come up when you use scrap material. Some questions come to mind *before* one uses the material, some *while* using the material and some *after* the material has been used. You might also want to look at the curriculum ideas involved.

Games

As a child I used to run up and down stairs. You can make up many different problems with stairs! Here is one: You can go up a certain staircase 2 at a time and have 1 stair left and go down 3 at a time and have 1 stair left. Does this determine the number of stairs? What if you are told that I can jump down 5 at a time and then there are none left over? Does that determine the number of stairs? If not, what else could you be told that would? You can make up a lot of different problems while you jump up and down stairs!

Noughts and crosses, known to Americans as tic-tac-toe, can give rise to a lot of problems:

- How many different starting points are there?
- What shall we decide is meant by different?
- What are some strategies to win?
- How can you record information without drawing pictures?
- What if you had a 4 by 4 grid?
- What if you played noughts and crosses in 3 dimensions?

Dominoes lend themselves to many problems. Children do a lot of counting while skipping rope. There are problems about one child double jumping every third beat and another every fifth beat. It was the potato race that made me think a lot when I was young. Recall that you have to pick up the potatoes, placed in a row, one by one and place them in a bucket before running to the finish line. I used to really wonder whether it was better to first pick up the one nearest or the one furthest away! Or would the middle one be best? Then there was also psychology involved as you watched others and saw how many potatoes were left! If I had 3 left — near me but the person running next to me had only 1 left further away ....

You might want to alter the potato race. Let's list some attributes — 5 potatoes, one at a time, all in one line, at equal distances... What if you had, say, ten potatoes, not all one by one, not all at same distance from each other — what kind of race can you make up?

And then there was the three legged race. If there are 24 'legs' running how many people are in the race? If there are 48 people running how many legs will be running?

I recall in the game of Tag where one could be safe by standing in front of a pair of players and the one outside would be IT. I recall having a distinct feeling for when something is added and the same amount is taken away that the result is the same.

Perhaps it is worthwhile to think back on the games you played and to consider what mathematical ideas may be lurking in them.

Card tricks of course too can be used to do a lot of maths. I have purposefully not mentioned games that have obvious maths in them.

Mistakes

If we have economy and we have little material there is one thing we will always still have lots of and that is *mistakes*. I will mention them partly because I hope that the person who has done much work with mistakes, Raffaeli Borasi, will publish some of her work in the English journals. She has had several other articles in *Mathematics Teaching*. Some of her articles about mistakes appear in [3]; they show lots of ways of using mistakes.

L. Myerson had an article in Mathematics Teaching [13] as did Tim Barclay. Let me first briefly mention

Tim Barclay's article called *Buggy*[2]. His idea was to present children with a common mistake such as the one shown below and show the same mistake in different examples. He then asked the children to find what the bug is and explain it. I use his idea a lot with my teachers in training.

$$\begin{array}{r} 4\ 3 \\ \times\quad 8 \\ \hline 3\ 2\ 4 \end{array} \qquad \begin{array}{r} 9\ 3 \\ \times\quad 6 \\ \hline 5\ 4\ 8 \end{array}$$

Extending the problem — don't always keep to special cases.

We spend a lot of time in the school curriculum on special topics without ever telling the students how special the cases are. For example, we spend a lot of time in geometry on congruent triangles. Do we bring up the question of why triangles may in fact be more important to study than quadrilaterals? Do we raise the question of what congruency conditions for quadrilaterals would be? [15], [18]. Perhaps my favourite example is one which I brought to the ATM some years ago. The problem arose when I was making plaster casts of geometric shapes with youngsters and we were making moulds for right circular cones. When the students had made some, I suggested they now try and make some moulds for a non right circular cone — an oblique cone with a circular base (not a cone obtained by slicing a right circular cone at an angle). They knew that they should draw the plan. Luckily the bell rang. I spent a lot of time trying to do it and then took the problem to the ATM meeting.

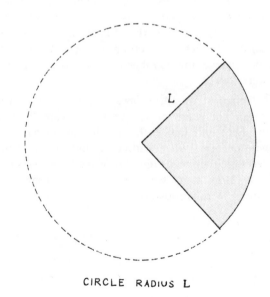

CIRCLE RADIUS L

The shaded part is a plan of the right circular cone overpage.

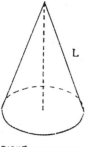 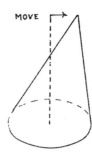

RIGHT CIRCULAR
CONE

MOVE

distance from front of first chair plus leg room to back of last chair?

○ *Some are back to front. How many similar arrangements can be made?*

○ *Five people sit on the chairs. In how many ways can they be seated so that they do not sit next to the same person twice?*

○ *Five people on chairs, tallest at front, smallest at back. How many changes are needed to reorganise so that smallest is at front, tallest at back?*

○ *Turn the chairs so that they all face the other way — but you have to turn 2,3,4,... at a time. What is possible? What not?*

A little change of distance for the vertex alters the plan drastically. Bill Brookes sent me a solution... It is not an elementary problem but it, too, is an example where we don't tell children how special a situation is. Surely we want children to wonder what happens if ... or what happens if not ...? You might be on the look out for other special situations.

Well there are many other situations from which you can get problems from real situations; for example when next you see stacked oranges you may ask yourself why are they stacked the way they are — how many on the next layer and so on. [16] [17] And I won't have time to talk about Maths from people — Nuffield did a lot of that.

I will end up by saying that if you have no books and no stuff, don't worry too much because you will do more thinking, more talking, more doing, more discussing, more inventing, more comparing, and have more fun but of course you will still have 118 issues of *Mathematics Teaching!* ∎

University of Oregon, Eugene

A great many people in the audience took up Marion's challenges to do some mathematics. First, thank you to all those who worked on the bottles problem, and congratulations to the winner.

Second, there was an enormous response to the chair's challenge (see MT120 p6) which resulted in many rich and interesting questions being compiled for use at all levels. We are able to include a very small selection which Marion chose to illustrate this diversity. Why is it that textbook questions, on whatever level, are so uninspiring by comparison?

○ *How many legs?*

○ *How many different positions for the gap?*

○ *Helen is sitting on the first chair, Edward is sitting on the last chair. How many moves must Helen make to reach Edward?*

○ *If you had to stack chairs in twos, how many stacks could you make? What about stacking in threes?*

○ *We need leg room to sit comfortably. What is the total*

References
1 Banwell C S, Saunders K D, Tahta D G, *Starting Points*, 2nd Edn., Tarquin 1986
2 Barclay Tim, 'Buggy', *Mathematics Teaching* 92, pp10–12
3 Borasi Raffaella, 'Using Errors as Springboards for the Learning of Mathematics', *FOCUS On Learning Problems in Mathematics*, Special Issue, Issue Editors Stephen I Brown and Leroy G Callahan, Summer/Fall Edition 1985 Vol 7 Numbers 3 and 4
4 Brown S I and Walter M, *The Art of Problem Posing*, Lawrence Erlbaum Associates, NJ 1983
5 Carman Robert A, ' Mathematical Misteaks', *Mathematics Teacher*, Feb 1971, p109
6 Carmony Lowell, 'Adding Fractions Incorrectly', *Mathematics Teacher*, Dec 1978, pp737-38
7 Coxeter and Greizer, *Geometry Revisited*, pp89–91, Math. Association of America, 1967
8 Dearborn R G, 'Integral Sides', *Mathematics Teacher*, Apr 86, p238
9 *DIME* Materials, available from Tarquin
10 Hargreaves M and Hughes Y, 10^2, AN ATM activity book, 1986
11 Jordan J H et al 'Triangles with Integer Sides', *American mathematical Monthly*, Vol 86 No.8, 1979
12 Leapfrogs *Imaginings*, EARO
13 Meyerson L, 'Mathematical Mistakes' *Mathematics Teaching* 76, Sept 1976, pp38-40
14 Pedersen Jean and Polya George, 'On Problems with Solutions Attainable in More than One Way' *College Mathematics Journal*, Vol.15 No.3 June 1984
15 Ranucci Ernest R, 'The Congruency of Quadrilaterals' *Mathematics Teaching* 64, Sept 1973, pp35-37
16 —, 'Fruitful Mathematics' *Mathematics Teacher*, Jan 1974, pp5-14
17 Sachs J M, 'A Comment on "Fruitful Mathematics"' *Mathematics Teacher*, Dec 1974, pp701-73
18 Spitler Gail and Weinstein Marian, 'Congruence Extended: a setting for activity in geometry' *Mathematics Teacher*, Jan 1976, pp18-21
19 Walter M, 'Polyominoes, Milk Cartons and Groups' *Mathematics Teaching*, No.43 Summer 1968 pp.12–19
20 —, 'A second example of informal geometry: Milk cartons' *The Arithmetic Teacher*, Vol XVI No.5 May 1969 pp368–370
21 —, *Boxes Squares and Other Things*, National Council of Teachers of Mathematics, Resten Va. 1971
22 —, 'Two Problems from a Triangle' *Mathematics Teaching*, March 1976 p.38
23 —,'Do we rob students of a chance to learn?' *For the Learning of Mathematics*, Vol.1 No.3 1981
24 —, 'Mathematizing with a piece of paper', *Mathematics Teaching* 93, Dec 1980, pp27-30
25 —, 'Exploring a Rectangle Problem' *Mathematics Magazine*, Vol.54 No.3 May 1981 pp131–134
26 —, 'The Day all Textbooks Disappeared' *Mathematics Teaching*, No.112 Sept 1985 pp8–11
27 *Geometric Images*, ATM 1982

CURRICULUM TOPICS THROUGH PROBLEM POSING

Marion Walter

Parts of this article include material presented at the 1989 ATM Easter Course at St Martin's, Lancaster — Eds

Problem *solving* has received much attention in the past years and younger children are beginning to be seriously involved in it. Problem *posing* is beginning to receive attention and it, too, can begin in the earlier years. Furthermore, if we encourage students to engage in problem posing, we can involve them more deeply in the development of topics that we wish *to cover*; in fact, we can use problem posing to help students *uncover* mathematics.

Addition problems

Students are often presented with pages of addition problems. What message does this give to the students? Surely one message is that they are not capable of making up their own practice problems. (I realise of course the advantage of giving students all the same problems — it is easier to check the answers!) What might one do instead and why?

Suppose the students are just beginning to learn how to add three digit numbers and that they have just worked out:

```
  3 4 2
+ 5 3 4
-------
```

where there is no need to regroup or 'carry'. If we ask them to make up more such exercises, they will soon be faced with the challenge of what numbers to choose so that the total in each column is less than 10 if they wish to avoid having to regroup. Surely this would be a learning situation.

Or, suppose they are given:

```
  3 4 2
+ 2 1 8
-------
```

might it not be worthwhile to challenge the students to make up some more problems where the units column adds up to ten?

Or, let's turn the task around. Given:

```
  3 4 2
+ 5 3 4
-------
```

what problems can you or your students pose?

One technique of problem posing just asks you to look at the given, in this case an addition problem, and asks you or your students to try to think of other problems. Stephen Brown and I have called this *accepting the given* [2], and we sometimes call it *brute force* problem posing. Among the suggestions from participants at Lancaster, using only this technique of problem posing, were:

— Here the answer is 876. Make up other addition problems of two 3-digit numbers whose sum is 876. What do you think students will be discovering or learning if they do this?

— Make up other 3-digit addition problems whose answers consist of three consecutive digits.

— Rearrange the digits of each 3-digit number in 342 + 534 to get the largest possible total.

— Find all the different totals you can get by rearranging the digits.

— Make up a story that goes with 342 + 534.

— Make up other 3-digit problems for which the answer is such that the digit in the tens place is one greater than the digit in the units place.

Notice that each of these problems can raise or lead to other problems. For example, the second raises the additional problem: *what are all possible 3-digit numbers that consist of consecutive integers?* And what about the totals of 876 and 678? Can one always or ever reverse the digits of the addends to get the reversed total?

```
    3 1 2            2 1 3
  + 5 6 4  and     + 4 6 5
  -------           -------
    8 7 6             6 7 8
```

```
but   3 5 9     9 5 3     1 3 7     7 3 1
    + 5 1 7   + 7 1 5   + 7 3 9   + 9 3 7
    -------   -------   -------   -------
      8 7 6   1 6 6 8     8 7 6   1 6 6 8
```

If it is not possible, will one always get 1668 instead of 678? Explore.

The fourth problem might give rise to the question: *how many different totals are there?*

You will think of many other problems even without using any other techniques of problem posing. In this way students can be engaged in problem posing and problem solving while also getting practice in addition. Students will be thinking and will be involved in creating their own problems. Thus at an early age they can experience some mathematics in the making. They can learn from experience that mathematics is *not* a subject in which you have to be told everything and memorise a lot.

Fractions

Next, I suggested a fraction excercise as a starting point:*what problems can you think of when faced with* $\frac{2}{3} + \frac{1}{5}$?

Usually students are given such problems, and they either have learned the algorithm for finding the answer or they make mistakes. They are usually not asked to think.

Among some interesting problems that were posed at Lancaster were these two:
— When in real life would you ever have to add these two fractions?
— How many different ways can you add these?

Some questions one might pose and which I imposed on the group are:
— Which is bigger, $\frac{2}{3}$ or $\frac{1}{5}$?
— Is the answer less or more than 1?
— By how much does $\frac{2}{3}$ differ from 1?
— By how much does $\frac{1}{5}$ differ from 1?
— What must one add to $\frac{2}{3} + \frac{1}{5}$ to obtain a total of 1?

I had worked this last question out and found: $\frac{2}{3} + \frac{1}{5} + \frac{2}{15} = 1$.
Using Polya's admonition, which he so often said in his class: *Look at the problem,* I noticed that one could get the answer $\frac{2}{15}$ by multiplying the numerators and denominators of $\frac{2}{3}$ and $\frac{1}{5}$. I had chosen $\frac{2}{3} + \frac{1}{5}$ at random when I wrote it down on an overhead some time before Lancaster and had calculated the $\frac{2}{15}$. I was lucky! This immediately raised the question: *What other fractions could one start with so that one could find the right answer by this 'wrong' method?* This is such a rich problem that I have since 'milked' it a great deal [1]

Geometric Figure

Next I took a geometric starting point: a regular hexagon.

Instead of beginning with what one wants 'to teach' about a regular hexagon, (what does one want students to know about a regular hexagon anyway?), let us brute force problem pose. Here are some of the problems and questions that were suggested by the group:
— How many diagonals does it have?
— How many triangles can you make?

Note that this is a good example of a question that would need to be clarified by students. This is a valuable activity because it is misleading to always give clearly stated and well defined problems and

learning to clarify problems is in itself a worthwhile activity.
— When all the diagonals are drawn in how many regions are formed?
— What is the length of each diagonal? Or one might ask how many different lengths are there?
— What is the area of the hexagon formed by joining the alternate vertices?
— What is the largest circle you can draw in it?
— If it is not rigid, what shapes can you deform it into?

I suggested one of my favourite questions: *can you construct a regular hexagon, not only by using straightedge and compass, but from a paper circle, or a rectangle, or a scrap of paper or an equilateral triangle?* [4].

If we draw diagrams suggested by some of the questions above, or other simple diagrams using a regular hexagon as a starting point, many more questions are suggested. Perhaps you want to pose a few now:

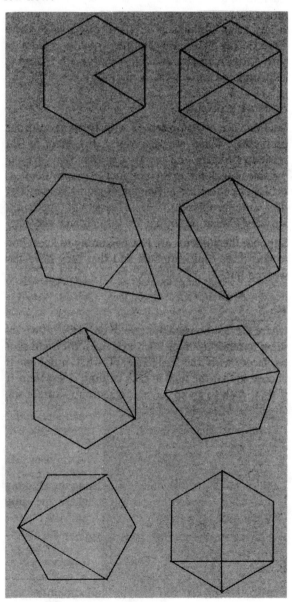

Other starting points

We briefly examined $3x = 12$ as a starting point [5], as well as the problem of how many different triangles one could construct out of a stick of length 10 cm if each side has to be a whole number of centimetres and the stick gets used up each time one triangle is made. This is a nice problem that really brings home the triangle inequality — the sum of two sides of a triangle is greater than the third side. Warning: No easy formula exists that tells you how many triangles are possible for sticks of length n cm [3].

Group work

Part of the time was spent by participants in small groups choosing their own starting points — ones useful for the curriculum items they have 'to cover'. Since there were teachers of all levels from infant to upper secondary, we got a variety of topics, ranging from a delightfully rich one that started with the dustbin to a trigonometry one: $\sin^2 x + \cos^2 x = 1$.

The group discussing the dustbin created many wonderful questions including:

— How many dustbins are needed to collect the week's school litter?
— How high is it?
— What else can you measure about this dustbin?
— How many ways can you fit the lid on the bin?
— Is it better to have two little bins or one big one?
— What do you think would fit in this bin — an elephant? a bird ...?
— Find the volume of the dustbin.

— If the bin is half full of water how high is the water level?
— How much bigger is the top rim than the bottom one?
— What else can you use if for? Would it be useful for storing toys?
— What if you could not reach into the bottom?
— What else can you put the rubbish in — what are some attributes of a good rubbish holder?

I believe the teacher who suggested this starting point had done a whole chunk of curriculum around the rubbish bin! After listing several questions suggested by $\sin^2 x + \cos^2 x = 1$, I pondered on how often we 'teach' this without expressing the wonder about how special it is. Can we say anything about $\sin(x) + \cos(x)$, for example, or $\sin^2 x - \cos^2 x$ or $\tan(x) + \sin(x)$?

I do not have a record of the work of the other groups at Lancaster, but I hope that even this small report will encourage others to engage in problem posing while trying to 'cover' the curriculum. ∎

Mathematics Department, University of Oregon, USA

References
1 R Borasi, Algebraic explorations of the error 16/64 = 1/4, *Mathematics Teacher*, April 1986, pp 246-248
  Wrong methods that give the right answers have been discussed in many places by several authors. R. Borasi has written much about it. I am not sure yet where the further discussion of the problem will be published.
2 S Brown and M Walter, *The art of problem posing*, Lawrence Erlbaum Associates Inc, Hillsdale, NJ 1983
3 J H Jordan et al, Triangles with integer sides *American mathematical monthly*, Vol 86, No.8, October 1979, pp 686-9,
4 M Walter, Are we robbing our students of a chance to learn, *For the learning of mathematics*, Vol 1, No.3, March 1981
5 M Walter, Some roles of problem posing in the learning of mathematics, chapter in *Mathematics, teachers and children*, D. Pimm (Ed), Hodder and Stoughton, London, 1988, pp190-200.

Below are two images from Budapest taken by Marion whilst at ICMI 6 − Eds.

SQUARES 5 - 19

This is the first of a series of MT pull-outs for the classroom. This pull-out is based on an account of work done by pupils and teachers in Cornwall, collected for Mathematics Teaching by **Jane Anderson**. Most of the following text is taken from an article called 'Continuity ... squares 5-19' by Jackie Long and Jane Anderson, first published in *Vector No. 16*. Vector is a mathematical newsletter for secondary schools in Cornwall.

At the 1985 ATM Easter Course Marion Walter gave the closing lecture, taking squares as her classroom theme. Throughout her lecture there were slides and numerous OHP transparencies which were accompanied by a flow of ideas and questions which left listeners mentally breathless.

The notes for her lecture were printed in MT112 the following September and have been used within Cornwall, first at an ATM Camborne Branch meeting when teachers worked on them, then with able Y10 pupils, and more recently with young infants. The infant teacher remarked that nearly all of Marion's ideas were usable with infants; the ideas could also stretch A-level students.

> The following two activities taken from Marion's article were carried out with a group of Y1 children aged between 5 years 5 months and 5 years 9 months. The work formed part of a project on Post Offices.

Activity 1: Making parcels

The children had to find as many ways as they could to arrange squares into a pattern in which they touched along an edge. They had a choice of Polydrons or ATM squares to use. There were only enough for 6 squares each so the children had to devise a way of recording each pattern so that they could re-use the squares.

The children had previously used 2cm squared paper to record multilink patterns and on this occasion they chose to use it again. All but one of the children were able to use one square on the paper to represent 1 Polydron or ATM MAT, in spite of the fact that the squares on the paper were of a different size to the physical objects.

When the children had made several arrangements they were asked to see which pattern would fold into a square box with a lid.

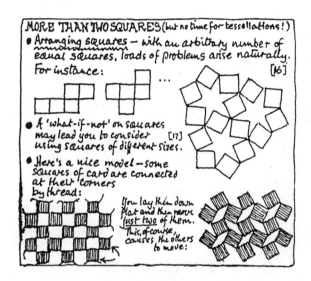

The children with the Polydrons found this very easy since they could fold up the polydrons and see if it would work. The children then noted beside their diagrams which would make a cube and which would not.

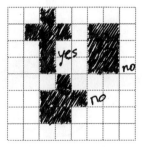

The children using the ATM MATs could not fold up the squares to make a box so they had to use other strategies. One child borrowed some Polydrons from another group. Talan (5 years 6 months) was able to visualise which of his patterns would fold and which would not. He could see that if four squares were put together to form a larger square then it would not fold. He used his hands as a substitute for polydrons lifting them from the table to imagine how the squares would fit together.

SQUARES

GROWING SQUARES

Take some regularly increasing squares. Arrange them in different ways.

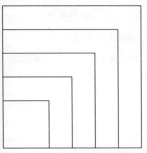 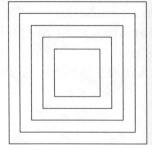

How wide are the steps? What is the area of the steps? Organise them in spirals. . .

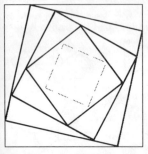 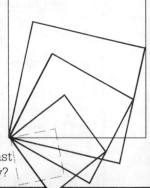

For these patterns the last square is too small. Why?

HALVING SQUARES

Take a square. Keep folding it in half to form triangles. Fit your triangles together in different ways.

Find out all you can about your pictures.

TOPPLING SQUARES

Topple a square round a rectangle. Does the corner marked with a dot get back to the same place?

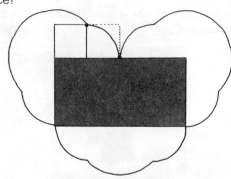

Plot the locus of the corner marked with a dot. What shape is it? How long is it? What area does it contain?

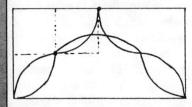

Try toppling the square inside rectangles.

ROTATING SQUARES

Rotate a square about its centre

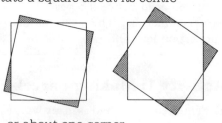

. . . or about one corner

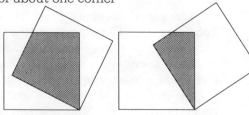

. . . or about the midpoint of one of its sides

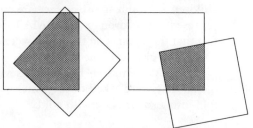

Find out about the shaded shapes.

SQUARING THE CIRCLE

Draw half or quarter circles inside squares. Shade your diagrams in different ways. What area have you shaded?

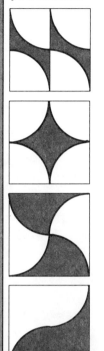

MIDPOINTS

Form a square inside a square by joining midpoints. Slide the inner square. What is the area of overlap?

Draw the diagonals of a square. Join two corners to the midpoints of the opposite sides. Discuss the triangles formed.

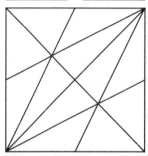

IMAGINE

What shapes are produced when

. . . a square rotates in space about one of its sides.

. . . a square rotates about one of its diagonals.

. . . a circle rotates about one of its diameters.

. . . a circle rotates about a line outside itself.

SQUARES IN SQUARES

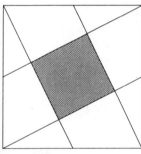

Convince yourself that the shaded shapes are squares.
How big are they?

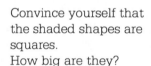
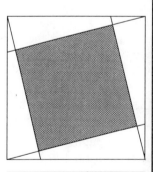

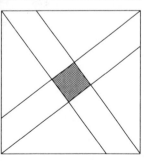

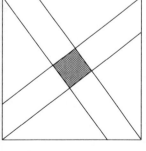

Find other ways of drawing squares inside squares.

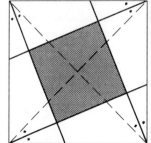

SQUARES IN TRIANGLES

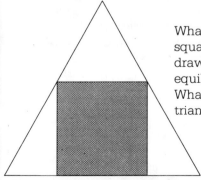

What is the biggest square you can draw inside an equilateral triangle? What about other triangles?

Activity 2:
Squares and curved areas

In order to design wrapping paper the children were given a square template and a circle template. They were asked to draw a square with a circle inside and then do the same with four small squares put together to make the next square and so on. The children were also asked to put a circle where the squares joined.

  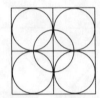

The drawing of these figures generated discussion about square numbers which I had not anticipated. The children tried to predict how many small squares would make the next big square. Having put four small squares together the prediction was that they would need six small squares to make the next big square. They tested this and found that they were wrong and so they thought about it. Some children predicted that they would need eight small squares and some that they would need nine. They tested their predictions to see which was correct.

When they had drawn the circles inside squares I wanted the children to generate some number patterns and so I asked them to count how many small squares were in each larger square and also how many curved areas the circles generated. The children then recorded this in a table.

This activity led on to some work which involved making squares with Multilink. The children were challenged to see what combinations of Multilink would make the squares and which would not. They recorded their models on 2cm squared paper and made a table of the numerical results. The 5 × 5 square challenged the children's counting ability and they had to refer to a large hundred square on the wall and count from 1 to see how to write their numbers!

GCSE coursework

The Y10 pupils were given a copy of Marion's notes, as printed in MT112 and told to choose one small part to work on in order to produce a piece of GCSE coursework. Initially, the pupils were overwhelmed and found it difficult to believe that they only needed to concentrate on one very small part in order to produce a good piece of coursework. Once they settled to working they began to realise that one small section really was very challenging and very rich.

What follows is a description of some of the work carried out by Daniel.

Daniel was motivated by the growing triangles, but grew his in a different way from that shown in MT112. He then decided that he would investigate the increases in the length of the hypotenuse. As the work progressed this became less of a practical activity and more an investigation into number patterns. He began by seeing two alternating patterns, viz 1.414..., 2, 2.828..., 4, 5.656... etc. which were the powers of 2 and 'doubles of the square root of two'. By considering the areas of the triangles he found that their areas grow by powers of 2. He then wanted to generalize; logarithms... fractional powers... teacher intervention was essential.

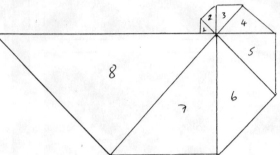

Triangle position	Hypotenuse length	No of Triangles in Triangle	Area
1	1·4142136 cm	1	0·5 cm²
2	2 cm	2	1 cm²
3	2·8284271 cm	4	2 cm²
4	4 cm	8	4 cm²
5	5·6568542 cm	16	8 cm²
6	8 cm	32	16 cm²
7	11·313708 cm	64	32 cm²
8	16 cm	128	64 cm²
9	22·627417 cm	256	128 cm²
10	32 cm	512	256 cm²

Resources

A useful resource for this work is a set of tissue squares and tissue circles. These can be obtained through most LEA suppliers. They come in regularly increasing sizes. The squares have sides of these lengths: 5 cm, 7.5 cm, 10 cm, 12.5 cm, 15 cm; and the circles have diameters of these same lengths. Tissue circles and squares make it easy to pursue some of the ideas shown, because the tissues can be superimposed and the interesting areas of overlap are easily seen. Apart from the tissue squares and circles, glue and sugar paper for mounting are all pupils need to produce pictures like the one on the front cover of this issue of MT. Children enjoy creating these pictures and this stimulates them to discuss some of the geometrical ideas contained within them.

ONE HUNDRED SQUARES REVISITED

Some mathematical activities focus on particular mathematical contents or are pitched at a particular level of difficulty. Bob Vertes describes an activity which is not specific in either of these ways.

I am writing this article to share an activity in which a number of mathematics learners and teachers have found pleasure and success. It has recently been used with teacher education students, whose attitudes and reactions to it have given me much food for thought.

As a PGCE course tutor I feel it is useful to remind secondary mathematics teachers that many children to whom they are about to teach mathematics neither recall not understand their 'times tables', even though they have done a lot of work on them in the earlier phases of their education. I also suggest to my students that what might seem very simple pieces of apparatus can lead to interesting mathematical development if explored openly.

One example of such apparatus is the one hundred square: something which my students know is used in primary schools, and which they often erroneously assume has little place in the secondary mathematics curriculum.

My usual starting point for a session with these students is to issue one hundred square boards and to ask students to place counters or cubes to cover the 2 times tables. I stop them after half a minute or so, to look round at everybody else's work: to notice how many are going 2, 4, 6, 8, 10 and then 12, 14, 16, ...; and how many are going 2, 12, 22, 32, 42 ... When stopped, some are about to switch from the first to the second approach; this is a good way of pointing out to graduates that there are different depths of understanding of and awareness of patterns in such a simple process, even among 'mathematicians' like themselves.

Asking students to cover the 3 times table on a cleared board then causes a few moments of hilarity, since there is no column pattern easily noticeable, or at least not until you have marked numbers to about 40 or 50; but a diagonal pattern seems more readily observable (figure 1).

1	2	**3**	4	5	**6**	7	8	**9**	10
11	**12**	13	14	**15**	16	17	**18**	19	20
21	22	23	**24**	25	26	**27**	28	29	**30**
31	32	**33**	34	35	**36**	37	38	**39**	40
41	**42**	43	44	**45**	46	47	**48**	49	50
51	52	53	**54**	55	56	**57**	58	59	**60**
61	62	**63**	64	65	**66**	67	68	**69**	70
71	**72**	73	74	**75**	76	77	**78**	79	80
81	82	83	**84**	85	86	**87**	88	89	**90**
91	92	**93**	94	95	**96**	97	98	**99**	100

Figure 1

That is not to say that some students don't go 3, 33, 63, 93, 6, 36, 66, 96, 9, 39, 69, 99, and then 12, 42, 72 or 21, 51, 81, preferring to work vertically rather than horizontally. Most, however, will go 3, 6, 9, 12, 15, 18, 21, 24, 27, 30 and then fill the rest using bottom left to top right diagonals. Some use the chess move 2 along 1 down, perhaps as a check.

At this stage I ask them to add the 4 times table onto the board. It comes as a surprise to some that the numbers with two pieces on are those in the 12 times table. Remember that these are graduate mathematicians, or people with a strong element of mathematics in their degrees. Many of them see bits of mathematics as so easy and obvious that they lack the sensitivity required to appreciate where pupils need time, encouragement and a practical approach. The simple activity above seems to be effective in bringing about a realisation that some re-thinking of their attitudes might be required.

I then ask them to put the 7 times table onto a blank 100 square printed sheet. I ask them to say how they did it: we compare approaches. Very few will use the same method for marking paper that they used for covering the board with cubes: most will have reverted to the 7, 14, 21, 28, 35 routine and

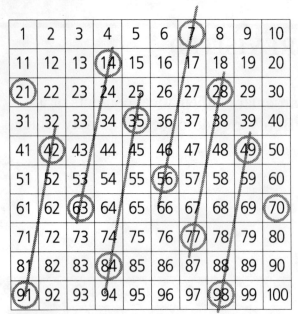

Figure 2a

Figure 2b

Figure 3

will have used it almost all the way through before spotting some sort of diagonal pattern to help them complete the task.

I then try to get students to write down in their own words how the table pattern of 7s can be most easily produced. The most likely response will be one of a small number of sets of 'diagonals' which the student sees in the pattern. As an alternative to seeing the diagonals a minority of students use a vector approach. They rarely say they are using vectors, but instead use terms like 'knight's move' or 4 along and 1 down. It is very rare, in my experience, that students actually recognise that they are using vectors: could this be because they are reluctant to see vectors arising from the seemingly simple hundred square?

[The topic of vectors is among those least understood or liked: certainly in A-level, but perhaps even in GCSE. Our undergraduates seem to find it one of the hardest topics to adjust to in the first year of their course.]

I now ask the students to draw one set of parallel diagonals through the points (the 7 times table) they have marked. By comparing notes or diagrams it is clear that there are quite a few variations possible, some of which are shown in figures 2a and 2b. I ask them to draw a second set of parallel diagonals through the multiples of 7 so that they produce parallelograms all of whose vertices are multiples of 7. They are asked to do this so that there are no multiples of 7 inside any of these parallelograms. Some possible parallelograms are shown in figure 3.

A little brainstorming follows, where students recognise that there are several answers to the problem, none better than the rest. Some then challenge themselves to find all the parallelogram shapes that can be created.

I now ask about the mathematics contained in their diagrams. From some come comments about the lengths of sides of the parallelograms, and others talk of perimeters – these suggestions lead to discussion and use of Pythagoras' theorem. One rewarding investigation I have undertaken concerns a possible connection between the areas of these parallelograms.

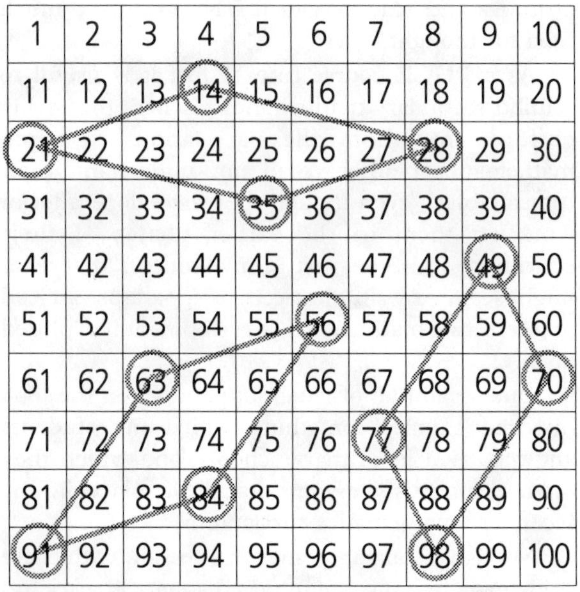

This activity produces useful mathematical discussion, since for each of the parallelograms that can be produced there might well be more than one way of calculating its area. If there is a right angle the process is particularly simple, but for one of the most commonly chosen parallelograms, with vertices at 7, 14, 35 and 28, there are two approaches worth considering.

The first attempt for most students is to try and find the area by breaking it into rectangles or right-angled triangles. This is not always possible, but it is possible here. The usual approach is to try and draw lines parallel and perpendicular to the sides of the

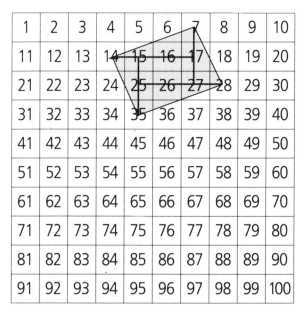

Figure 4

1	2	3	4	5	6	7	8	9	10
11	12	13	14	15	16	17	18	19	20
21	22	23	24	25	26	27	28	29	30
31	32	33	34	35	36	37	38	39	40
41	42	43	44	45	46	47	48	49	50
51	52	53	54	55	56	57	58	59	60
61	62	63	64	65	66	67	68	69	70
71	72	73	74	75	76	77	78	79	80
81	82	83	84	85	86	87	88	89	90
91	92	93	94	95	96	97	98	99	100

Figure 5

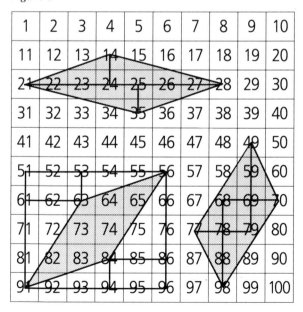

Figure 6

100 square, in this case creating two pairs of right-angled triangles and a 2 by 1 rectangle (figure 4). This dissection method is for many students the first or only routine attempted; this reliance on a single method can cause difficulties with some parallelograms.

The 'enclosure method' (figure 5), putting the shape into a rectangle and then finding the required area by discounting the wastage, is often as easy, sometimes easier and occasionally the only way. It is worth recounting that some of my PGCE students are so reluctant to use this method that, if the dissection method doesn't work, they find the angle in one corner using a protractor, find the side lengths of the parallelogram using Pythagoras' theorem and use the $\frac{1}{2}$ absin C formula!

In figure 6 there is displayed a range of approaches to finding the areas of three of the many possible parallelograms. Whichever method is used the area of parallelogram is 7 square units.

Not only do all the parallelograms so far referred to have an area of 7, but so do all possible correct parallelograms that the students may have drawn!

Then someone remembers that we have been using the 7 times table. Someone will ask if this is a coincidence. At this point the group will, if allowed or encouraged, start looking at, for example, the 6 or 8 times table, discovering that the corresponding parallelograms have areas of 6 or 8 square units!

Now for some this comes as a great surprise; for most the result causes a little bit of pleasure. For some (I am not numbered among these) it seems obvious, and yet I know of no proof that convinces me why it should be so.

I was delighted when Gillian Hatch, of Manchester Polytechnic, suggested extending the problem to see what happens if we use not the traditional hundred square but a different layout of numbers – for example with twelve numbers in a row instead of ten. I leave you to discover the effect on the area result.

I have used this activity on many different occasions and it seems suitable for 'top juniors', for early and older secondary groups, for PGCE students, ATM workshops and INSET sessions in schools.

I am indebted to an article in Mathematics Teaching [1] and to the booklet and poster which ATM later published [2].

Bob Vertes teaches at St Mary's College of HE, Strawberry Hill, Twickenham.

References

1 One hundred things to do with a hundred square, MT100
2 10^2, ATM, 1986

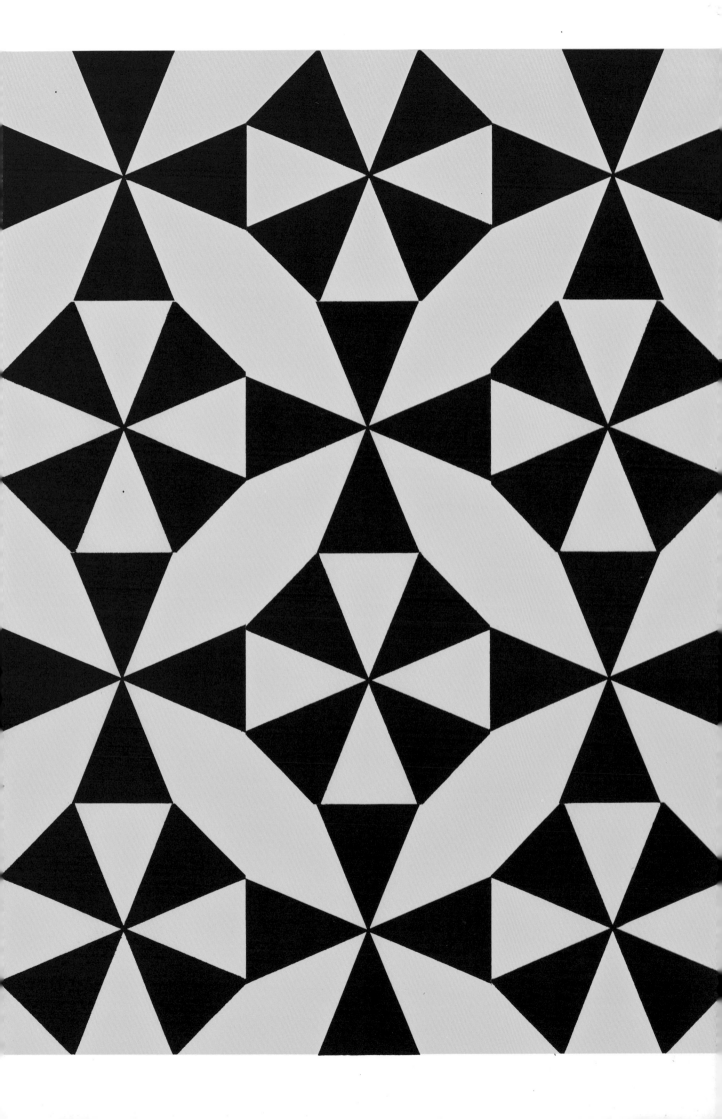

Quilting Tiles. Design made up using ATM 'Quilting Tiles', devised by Jenny Murray.

SECTION 3: DEBATES AND ISSUES

Debates and Issues

The writings in this section focus on some of the wider issues that teachers need to consider in their classroom. The first article, to which a number of writers contribute, is taken from Micromath and explores some issues related to equal opportunities. This is done with particular reference to gender and the use of technology within the classroom. Various strategies are discussed and will encourage the reader to be reflective about his/her actions in the classroom. Leone Burton and Ruth Townsend's article on Girl-friendly Mathematics explores the issues of gender in more depth and invites the reader to consider ways of fostering positive images of Mathematics with girls. Supporting students in their learning of Mathematics is clearly a fundamentally important role for all teachers, and different students require different kinds of support. Jim Smith's article focuses on the way support could be offered to students who are perceived to be particularly gifted in Mathematics.

Mathematics is often considered to be a neutral subject. Jenny Maxwell's article questions this assumption and uses a variety of examples from around the world to illustrate her thesis that mathematics does play a role in furthering social causes and political understanding. Linked with these ideas are the series of articles by Sharan-Jeet Shan and Peter Bailey. These are concerned with exploring the cultural context of Mathematics. They ask us to consider whether we are merely trying to multiculturalise our Mathematics or whether we are concerned to use Mathematics as a means of raising issues of equality and justice. There are many powerful and thought-provoking words here as well as ideas that could be used in the classroom.

The use of technology in the classroom is a debate which continues to arouse strong feelings in many people. Two articles here will help you explore some of the issues involved. In the first, Richard Noss considers the development in.the production of software for use by children. In particular he considers the move away from short programmes with limited flexibility and potential to the use of computer environments which allow students to develop some control over their learning and development. The second article by Barbara Hepburn considers the use of calculators with young children and how parents need to be involved with their children in understanding the learning potential within the calculator environment.

The final series of articles tackles the thorny issue of the assessment of Mathematics within the school context.

Tandy Clausen uses her experience of working in a special school to question the dominance of written tests as the medium for assessing student progress in Mathematics. She advocates more adventurous thinking and in particular the use of practical tasks as a means of assessing what students can do and understand. This is followed by Mike Ollerton's article in which he uses his experience of the ATM 100% coursework GCSE examination to consider a holistic approach to the assessment of Mathematics as an alternative to the "bits and pieces" approach of standard written tests.

Shirley Waghorn describes one strategy she has used to address gender issues in the classroom. *Glyn Holt* gives a considered reaction to the underlying principles. and *Maria Goulding. Pamela Greenhough. Martin Hughes. Katrina Laing* and *Teresa Smart* respond.

"*Debate*"

A Strategy for the Classroom

Shirley Waghorn

At an inset day on raising awareness on gender issues it was suggested that teachers were much more likely to ask a boy to answer a question or give an opinion than a girl, and that this was the case in both situations where the teacher selected a child to answer (regardless of whether the child had signified the desire to respond or not) and when the teacher had asked for 'hands up'. Whatever the situation, boys tend to draw attention to themselves, eg by sitting up very straight and pushing the upper body forward or by waving the arm directly at the teacher. When girls behave in a similar manner they are more likely to be told 'not to make an exhibition of themselves' than to be asked to answer the question!

I, personally, felt that such things did not occur in my classroom - boys and girls were treated equally - but decided to monitor certain areas to provide concrete evidence to belie such statements as those made above. However, to my consternation, I found that I was 75% more likely to ask a boy to respond than a girl (regardless of the subject matter) and, worse still, in any technical subjects such as science, the percentage increased in favour of the boys.

I decided positive action was required and adopted a policy of always asking a girl first and then alternating boy, girl to respond to questions. A colleague determined that he would give girls equal opportunities, at least with regard to using the computer and he adopted a policy of pairing the children in like sexes. Two girls were then always given first opportunity to use any new software, after which pairs of boys and girls took it in turn to work on the computer. To ensure that girls had equal opportunities to build with Lego, another colleague decided to set aside specific times when only girls were allowed to use the Lego bricks.

The policies adopted may not be the best way of tackling the problem but I feel that some positive action is better than no action!

Managing the Equality of Opportunity

Glyn Holt

"I'm sorry – I think I'd like to hear what Karen has to say on this issue." Two desperately disappointed people at a stroke – the one, David, cut off by the teacher in mid-delivery having recently learnt the power of the spoken word and the enjoyment of speaking to an audience; the other, Karen, who would have been quite content to make her own quiet assessment of the debate is now thrust forward to be scrutinised by the group.

"I'm sure we'd all like to see just what Sirath has achieved this morning." Sirath has been struggling to express himself within his second language, English, and his support teacher has just presented the word processed print-out to the teacher. Sirath is not sure whether *he* wants everyone to see, although he would have been really pleased to take it home to granny!

I witnessed both these events in schools where I have worked. They give me cause to reflect on two issues. The first is that everyone should have their right to silence respected, without being judged a non-participant. The second is that when a child attains a personal goal this is a private triumph and not necessarily for public consumption. Both transactions lacked two essential elements, negotiation and consent.

I have never been comfortable with 'positive discrimination'. I think Lord Justice Scarman first used this phrase back in the '80s during his inquiry into the so-called Race Riots in Brixton . Scarman judged Brixton's black population to feel alienated within a predominantly white culture. Pressed by a need to survive, some members of this community had expressed themselves through violence. Scarman went on to recommend positive social and economic discrimination in order to redress the balance. My difficulty is that I feel the race issue, like that of gender, is not simply restricted to one particular generic group but part of the global issue of equal opportunities. In other words, to discriminate positively is nevertheless *still* to discriminate. It emphasises the differences between us all, thus denying the more fundamental similarities, the sheer human-ness, that we share.

I come from a background of teaching in the special needs sector (just note that, the special needs *sector* !). Historically those whom society perceives as different are dealt with by being artificially stereotyped as a sub-group. Sometimes this is with the best of intentions and sometimes the very worst. Our humanity abhors the excesses of genocide or sectarian violence but often glides over the alienation that we apply at the other end of this continuum. I have spent over twenty years of my working life in schools *set aside* for those with special needs, yet paradoxically I have always claimed that there is nothing special about special schools. In my view they exist fundamentally so that mainstream schools can avoid their responsibilities to provide equal opportunities for all children.

Race, gender and special educational needs are not separate issues and should not be dealt with separately. In fact they should not be dealt with at all, except as a small part of the larger 'entitlement' debate. I hope to continue, in my dealings with humanity, in recognising and respecting the differences between myself and others, and celebrating them. Where I need to put my energies is in sensitising myself to the obstacles that society puts in the way of individual fulfilment. I need to consider the subtle implications of one culture over-laying another. I should be aware of the underlying messages contained in the words I use to transact my daily life. I need to consider what my body language is saying when my verbal language is giving another message. I would be escaping my duty if I did not frequently check that the 'curriculum' I think I am delivering is the same one that my fellow human beings are

receiving, or at least that the gap between the two is as narrow as possible.

We now talk of an entitlement curriculum, differentiated so that everyone can play as effective a part as their abilities allow. That curriculum has a social as well as an intellectual element. Every working day we negotiate the terms of a learning experience with our pupils, whether consciously or not. It is inappropriate to ask girls more questions than boys to 'bring them out', or to have a learning support teacher working in a corner alongside a pupil, maintaining a sort of intellectual apartheid. The issues are more fundamental than applying simple mechanical solutions to sociological problems. They are part of the ethos of the school and of society, and need examining in a wider forum. They are part of the group's acceptance of its duty to consider the dynamics within itself. Individuals need to ask themselves constantly whether their actions exclude the individuality of others.

Imposition of control over group dynamics, whether by the teacher or dominant child, is always monumentally unproductive. The questions are not "Are the girls getting a look-in in this debate?", but "Is there *anyone* who might need support to express themselves?", not "What do I as a teacher need to control in order to combat discrimination?" but "What does this child need to achieve today?", " Have we discussed the issue and agreed a contract between us?"

There is another balance to redress, too. Consider David who, in the example above, had to give way to the unwilling Karen. In some quarters to be male is to be stereotyped as the oppressor, the stealer of opportunity, and the intellectual bully. Sensitive young people (David was such a child) and adults can often fail to deal with the damage caused by the sometimes sudden realisation of how society perceives their rôle. The suppressed aggression that is often directed at them may be too painful to handle with dignity. The apparent difficulty of rationalising their position can prove too burdensome, resulting in a withdrawal from group activity. What help do we extend to David? Do we even notice the pain he suffers, or do we perhaps see him as surly and uncooperative? How do we address his need for self-respect? It is no accident that men's movements have recently and spontaneously formed in response to an emerging need for reassurance, and to mend the damage caused by a sometimes too strident and self-focussed advocacy of women's rights. In truth there are no rights, only privileges. If we mistakenly take our privileges to be rights we take a step backwards.The way forward is by a consensus of the group, and the skills necessary to bring *this* about need a serious commitment of time and thought. Nevertheless they are skills that *can* be learnt like any others.

You can't rig the game simply by moving chairs about the classroom, separating cliques, splitting friendships or plugging in a spotlight and shining it on that unwilling and wide-eyed face. If we want to be positive about anything, we shouldn't be positive in our *discrimination*, but in our *affirmation* that everyone has individual needs which can be met as part of a whole community. If those needs can serve concurrently both the individual and the group, then maybe Sirath will one day be happy to show us the history of his family, lovingly set down in dot-matrix print, whilst David might chair a discussion with sensitivity and discretion, and Karen might champion herself and others in joyous debate.

Examining Assumptions

Maria Goulding

'Equal opportunities' is a phrase which means many different things to different people. Just as in the language of mathematics, where a word like *difference* can have a precise meaning which may be confused with its everyday use, so too with equal opportunities. For some, the phrase is to be identified with the policies designed to tackle the problem of particular groups of pupils underachieving in our educational system. For others it is a blanket term which describes the entitlement of free school education for children aged 5 to 16 regardless of outcome. Increasingly, however, in the official language of policy documents and government charters, equal opportunities has been narrowed down to mean women.

There is a danger, in my view, of isolating gender issues in this way or even of organising related issues along a continuum:

'race' and gender...disability...(and for the really daring) sexual orientation

It becomes increasingly apparent to me that the glue which holds the issues together is power, or lack of it. Yes, we are looking at entitlement and the celebration of difference as Glyn Holt argues, but there is much more to it than that. It is not an oversight on the part of teachers that has led to some children receiving preferential treatment in the classroom or computer room; a complex web of connected assumptions rooted in, amongst other things, teachers' beliefs and structures in schools may at worst perpetuate the power differentials which exist outside school. Does the continuum I have suggested above look the way it is because the groups along it are themselves in a pecking order of disadvantage or powerlessness? In treating issues discretely we may well be losing sight of fundamental and unifying principles.

So where does learning mathematics fit into all this? Gender issues have often been treated in isolation here, largely because of the perceived importance of the subject in the school curriculum and the perceived underachievement of girls [1]. To begin with, I would like to examine a well worn path. It is commonly assumed that boys dominate the discourse in the classroom. Redressing this imbalance and giving the girls a fairer deal seems a worthy aim for the teacher. However, simply asking more questions of the girls may be an inappropriate way of proceeding. They may not want to be singled out and their silence may not indicate a lack of understanding at all. The teacher may simply be reinforcing an existing value judgement – that vocal participation is a necessary prerequisite for success in learning mathematics. She may be assuming that girls have a problem and, in order to overcome it, they need to behave more like boys. Closer inspection may reveal a far more complex pattern of interaction, where differences within the behaviour of the girls or boys in a class are as marked as differences between them. What about the quiet boys and the vocal girls? Further and more importantly, it may not just be a question of looking at behaviour patterns. For instance, Valerie Walkerdine [2] demonstrated that *similar* behaviours in boys and girls resulted in different *interpretations* of their mathematical ability.

I am not suggesting that conscious management of discussion in the classroom is pointless, rather that directing more questions to girls is a simplistic strategy. It may be more fruitful to look at who holds the power, who needs empowering and how different learning styles can be harnessed to achieve this. Groups who have traditionally, and with some evidence, been identified as disadvantaged may benefit from this analysis but the approach can also be used when the classroom does not seem to fit other people's patterns. So the teacher may well find that it is largely, but not entirely, girls who are disadvantaged and can act accordingly, but she may also be using the same principles in a class where, say, the children of professional parents are getting more than their fair share of attention. One of the ways of handling talk in the classroom may now be to offer a range of different opportunities for talk besides working on the usual 'teacher - pupil - evaluation' model.

The consideration of learning styles leads us on very naturally to the use of technology, which should provide us with increased opportunities for flexibility. Again, a common assumption is that girls like to work cooperatively so that, for instance, using Logo to set and solve problems in small groups may be seen as a good way of building upon this preferred way of working. As a strategy this may have more in its favour than the directed questioning, but we do need to avoid seeing this strategy as a 'catch-all' easy solution. Research into group interaction with Logo [3] revealed a more complex picture than is often imagined. Here there were gender differences but it was not simply a question of cooperating or not:

> "girls *do* plan and boys *do* collaborate but not necessarily in the way we would predict, and who achieves more is at least partly dependent on how we organise and assess the activity."

Intriguingly, it also became evident that teachers' perceptions of the girls changed over the period of this longitudinal study. Looking through the lens of gender in this way may not produce simple and generalisable patterns but it does begin to throw light upon aspects of the teachers' work which have implications for equal opportunities in its widest sense.

So far I have been trying to unpick two common assumptions, not to devalue them. Looking for patterns may help us to find better ways of enpowering groups and individuals, but we need to try and reveal their limitations. We may find out more by focusing on class or gender differences, but our conclusions may inform situations where the power imbalance is not necessarily in those areas.

Further unpicking of these assumptions may also be useful in illustrating the two pathways which have been seen as ways forward in the equal opportunities debate. Directed questioning seems to belong to a range of strategies which have encouraged the powerless to adopt the values and behaviours of the powerful. In the case of girls it goes with a 'deficit' view of their learning which can be both patronising and counterproductive. However, since it is unrealistic to expect powerholders to give up their power without a struggle, it is a strategy which may have to be adopted, amongst others.

The second strategy, encouraging cooperative ways of working, belongs to the school of thought which embraces a gender inclusive curriculum. Here, rather than looking at girls as 'lacking' in some way, it is assumed that most of them have been socialised differently, but not in an inferior way, from many boys. Here we are trying to shift the values of a dominant group, not just change its membership.

A third strategy, for which technology can be a powerful tool, is to bring equal opportunities to the attention of pupils as a part of the mathematics curriculum. Here, pupils are actively encouraged to contextualise their mathematics by examining issues of social justice. Data handling provides the legitimisation, should it be needed.

For example, the excellent graph plotting program *Mouseplotter*[4] has global information collected together in a datafile called *Nation*. It is an extension of the file Glostat produced by the centre for Global Education at the University of York for use with *Quest*, and benefits from the sophistication and ease of use of the mouse driven program. My PCGE students in the past have been both staggered and absorbed at the access to such information and the ability to ask and answer questions which throw light on issues of inequality across nations. This could be part of work which also examines the selection and presentation of statistics in the press and looks for examples of misrepresentation not just in diagrams but in the body of the text where numbers and percentages are often slipped in to support arguments.

By actually raising such issues in mathematics classrooms this third strategy ranks alongside moves in the other two directions to work towards shifting the conventional power bases, changing them and redressing past imbalances.

In this article I have tried to explore equal opportunities issues in mathematics education by looking at power. As I visit schools and talk to students, it becomes increasingly obvious to me that the dominant organising principle of knowledge and pedagogy in mathematics education is not gender or race or sexual orientation. The construct of ability stares me in the face, and I am left with a big question. Does an analysis of power relations help me with this one and what strategies can be adopted to redress inequality here, given the present political climate.

References
[1] Ernest, P. (1991) *The Philosophy of Mathematics Education*, Falmer Press.
[2] Walkerdine, V. (1989) *Counting Girls Out*, Girls and Mathematics Unit, Institute of Education.
[3] Hoyles, C. and Sutherland, R (1989) *Logo Mathematics in the Classroom*, Routledge.
[4] *Mouseplotter*, The Shell Centre, University of Nottingham.

BOYS OR GIRLS: TOGETHER OR APART?

PAMELA GREENHOUGH,
MARTINA HUGHES AND
KATRINA LAING

Shirley Waghorn is right to prompt us to consider ways in which inequality based on gender may be constructed within the classroom. In our experience however, the situation is complex, and simple manipulations of classroom organisation may not necessarily solve the problem. This is particularly so if the assumptions underlying such manipulations are not subject to careful and critical examination.

The use of same-sex rather than mixed-sex pairs for computer work is suggested by Shirley Waghorn as one method for combating gender bias. Underlying this suggestion are two separate (although related) assumptions. The first is that mixed-sex pairs are likely to be dominated by boys - either physically or intellectually - to the detriment of girls. The second assumption is that girls are likely to learn more effectively in same sex pairs than in mixed-sex pairs. Our own research provides little support for either of these assumptions.

Over the last few years we have carried out a series of studies in which we have systematically compared children working in same-sex and mixed-sex pairs. All the studies have involved children aged 6-8 years using simple Logo commands to control the movements of a floor turtle. Most of the studies required the children to take the turtle around a specially constructed obstacle track, although in one study the children used the turtle's pen to copy shapes or construct their own designs. The children always worked in pairs of similar ability.

In none of these studies did we find any evidence of boys dominating girls in mixed-sex pairs. Where chairs were placed equidistant from the keyboard, boys did not crowd out the girls. Nor did they monopolise the keyboard and take the lion's share of the keypressing. In one typical study, for example, we found that out of ten mixed-sex pairs, girls were the dominant key pressers in five pairs, boys were dominant in four pairs, and in the remaining pair the key-pressing was shared on a 50-50 basis. This hardly provides convincing evidence of male domination.

Of course, it could be that the boys dominate mixed-sex pairs intellectually rather than physically - that they operate as the 'thinkists' whilst the girls function as the 'typists' [1]. However, we have found no evidence in our research for such a division of labour. In another typical study [2] we examined in some detail the social interaction surrounding a critical set of moves. In particular, we looked where the moves 'originated' - in other words, who's idea was it? In the mixed pairs we found that the girls originated 58% of the moves while the boys originated 42%. When there was disagree ment, the girl's move was entered on 57% of the occasions whilst the boy's suggestion was followed on 43% of occasions. In other words the girls were, if anything, the dominant party in this particular set of decisions.

Needless to say, our concern should not rest solely with classroom organisation and associated interaction processes. The question of outcomes is also of crucial importance. We need to consider how group composition and interaction influence learning and the development of attitudes.

In the first study we carried out in this area, we found that the girls who worked in same-sex pairs—the type of pairing advocated by Shirley Waghorn—in fact performed substantially worse than the girls who worked in mixed-sex pairs [3]. In subsequent work we have failed to replicate this particular finding. Nevertheless, we have not found in any study that the girls who worked in same-sex pairs perform significantly better than girls whose experience was in mixed pairs. Nor, indeed have we found any significant differences between boys who worked in either form of pairing. Thus from the point of view of learning outcomes, our research does not favour any particular form of pairing.

The picture for attitude change, however, is somewhat different. In one study, where the children's attitudes were monitored over time, the attitudes of boys who worked in mixed-sex pairs appeared to become less stereotyped. At the start of the study, half the boys in each group (same-sex or mixed-sex) rated girls and boys as equally competent in their work with the turtle, whilst the other half gave the boys a higher rating. At the end of the study, the opinions of boys from same sex pairs remained unchanged. However, all the boys from mixed-sex pairs rated boys and girls equally. In other words, if our concern is with changing children's attitudes, then mixed-sexed pairings may actually be preferable.

The relationship between interaction processes and learning outcomes is also raised in the contribution by Glyn Holt. In proposing that we respect "the child's right to silence" he is indirectly questioning the impor tance of talk for learning. This is a refreshing approach, given the generally uncritical acceptance given to the assumption that talk is desirable for learning. Again, our research studies on 6-8 year-olds using the Logo turtle throw some light on this assumption.

In one of our studies, the amount and nature of chil dren's talk during paired sessions was analysed and related to subsequent learning outcomes. Somewhat to our surprise, we were unable to find any positive correlations between aspects of talk and the children's individual performance; indeed, we actually found some negative correlations (similar findings have been reported by Webb, [4]. In other words, just because children are generating large amounts of apparently valuable talk, it does not necessarily mean they are learning what we want them to learn.

Moreover, it would appear that in group computer-based activities some children who contribute little to group discussion may learn a significant amount through watching the activities of others.

The findings of the research are by no means definitive or conclusive. They do not 'prove' that mixed-sex pairs are always better than same sex pairs, or that talk is irrelevant to learning. What our findings do show is that some of the assumptions frequently made in dis cussions of equal opportunities may not necessarily stand up to careful examination. Children's learning is a complex business, to which simplistic solutions may fail to do adequate justice.

References

[1] Sheingold, K., Hawkins, J., and Char, C. (1984) "I'm the thinkist, you're the typist": The interaction of technology and the social life of the classroom. Journal of Social Issues 40 (3).
[2] Greenhough, P. (1990) Girls, Boys and Turtles; An investigation using mixed and same-sex pairs. Unpublished M. Ed thesis, University of Exeter.
[3] Hughes, M., Brackenridge, A., Bibby, A. and Greenhough, P. (1988) Girls, Boys and Turtles: gender effects in young children learning with Logo. In C. Hoyles (ed) Gins and Computers, University of London Institute, Bedford Way Papers 34.
[4] Webb, N. (1984) Microcomputer learning in small groups: cognitive requirements and group processes. Journal of Educational Psychology 76 (6) 1076-1088.

A GRAPHIC BOOST
FOR GIRLS

TERESA SMART

"I've had my graphic calculator for six months but I haven't even taken it out of the box", admitted a teacher at the start of a graphic calculator workshop. "I saw no point in learning to use the graphic calculator because I couldn't see how I would use it in my class room. But my male colleagues immediately wanted to 'master' the graphic calculator by working through the manual."

Use of graphic calculators reveals some of the same gender differences as other calculators and computers, but when used carefully they can encourage collaborative work and mathematical thinking, and redress some of the differences between boys and girls.

Over the past three years I have been working with the graphic calculator in 11-16 mathematics classrooms. Although the graphic calculator is small and private, much involved mathematical discussion develops and collaborative work can be encouraged even if pupils work individually. For example, an introductory task is to investigate the role of m and c in the equation y = mx + c. After exploring straight line graphs, the pupils work through a puzzle sheet. Soon it becomes sensible for two or three pupils to collaborate. As they find out that a single wrong graph can't be rubbed out, one calculator becomes the 'try out' calculator and the other is used for the final version. Another task that leads to successful sharing and discussion is one that explores the set of quadrilaterals and their relation ships. The four vertices of a square can be plotted on the screen. The last point plotted is still live and, as such, can be moved around the screen using arrow keys. The aim of the task is to convert the square into other four-sided figures such as a rhombus, parallel ogram or trapezium. Students have found it is almost impossible to sit on their own wondering whether they really have a parallelogram. Pupils compare with a neighbour or argue in a group.

WHAT ABOUT THE GIRLS?

Most 11-16 pupils do not come across a graphic calculator before they meet one in the classroom. Girls are not yet disadvantaged. It is important that we learn from past experience as it is well documented that boys dominate the use of computers at school and at home [1].

The graphic calculator is private to the pupil and one mathematics head commented that "Girls in my school prefer working with the graphic calculator because of the privacy it affords. They can make mistakes without the fear of ridicule from the boy computer experts". This too has a contrasting side. One teacher in a graphic calculator workshop commented on her anxiety when she looked around to find all the participants working individually and it seemed "they all knew what to do, I felt excluded."

The recent NCET A-level project [2] showed that, in the group of students who had had access to graphic calculators as 'a standard mathematical tool', the girls performed better than the boys in subsequent tests, while in the control group who had not had regular access to graphic calculators, girls did not do as well as the boys.

Some lessons can be learned from experience with computers. Computer use in the classroom can encourage group work and discussion. Girls working on computers benefit from cooperation and are seriously undermined by competition. One study [3] looked at pupils using computers in a simulation task over a period of 10 days. The pupils worked in a cooperative group, a competitive group or in a group where they were assessed individually but were not in overt competition with other group members. The study found that girls did better when working in a cooperative group. In particular, "competition among students over who was most successful in the computer assisted instruction seemed to have an especially debilitating effect on the female students" and after the activity the girls in the competitive group "felt less confident in their ability to work with computers, liked computers less, liked geography less, and felt less supported personally and academically by the teacher".

I think it is important to encourage students to work collaboratively on the graphic calculator, while still enabling experimentation and mistake making in private.

VALUE AND PROCESS

The graphic calculator is an important resource for exploring mathematical ideas. A group of SMILE teachers writing about successful work with girls and computers in the maths classroom explain that "the computer is treated as just one other resource that pupils may need to use to achieve an understanding of the task they are working on"[4].

However, mastery of a graphic calculator can become a status symbol in itself. Recently I worked with a group of five boys over several sessions. They were very keen. Each day they wanted the session to continue and looked forward to the next. They were willing to collaborate and when one had difficulty, another would explain. However I was struck by the need they had to conquer the instrument. What else will it do ?" they asked. "I have done this; can we go onto something new?". The emphasis was on finding out about all the facilities—the product—rather than on using any one facility to explore an idea further— the process. Hoyles and Sutherland found that boys using Logo chose more well defined tasks that they could complete whereas the girls chose more loosely defined goals. They go on to "suggest that in our school culture, the more well-defined a goal is, the more value it is awarded.... Girls' apparent lack of emphasis on product can easily lead to their work being undervalued".

The value of a calculator is not only in enabling the student to plot graphs of whatever complexity easily and quickly, but to use this graph-plotting facility to generalise about families of curves and to be involved in mathematical activity. In my experience girls are ahead of the game in this. They will use one facility of the graphic calculator to explore ideas fully and only find out about another facility when they need it. This should be encouraged if we are to promote the learning of mathematics and give further support to girls. But the new emphasis on tests and the diminished role of course-work will probably work against this.

References

[1] Culley, L., (1988) Girls, Boys and Computers, Educational Studies, Vol. 14, No. 1.
[2] Ruthven, K., (1990), The influence of the graphic calculator use on translations from graphic to symbolic form, Educational Studies in Mathematics, Vol.21.
[3] Johnson, R. T., Johnson, D. W., Stanne, M. B., (1985), Effects of co-operative, competitive and individualistic goal structures on computer assisted instruction, in Computers in Education 5-11, ed. A. Jones, P. Scrimshaw, Open University Press, 1988.
[4] The SMILE Teachers, (1988), "My Mum Uses a Computer, Too", in Girls and Computers, Ed. C. Hoyles, Bedford Way Papers 34.
[5] Sutherland, R., Hoyles, C., Gender Perspectives on Logo Programming in the Mathematics Curriculum, in Girls and Computers, Ed. C. Hoyles, Bedford Way Papers 34.

Shirley Waghorn works at Sussex Schools Support systems in Brighton. Glyn Holt works for NCET. Maria Goulding works at Durham University. Pamela Greenhough, Martin Hughes and Katrina Laing work at Exeter University. Teresa Smart works at North London University

GIRL-FRIENDLY MATHEMATICS

Leone Burton and Ruth Townsend

This article is a report of an initiative to change the perspective which girls have of mathematics and its relevance to their future.

The format was a one-day conference: 'Be a Sumbody'. The first two such conferences to be run in the UK took place at the Open University in the summer of 1982. The responses of the girls, and of the teachers who accompanied them, were such that the organisers were convinced of the validity of running such events. One of the present authors (LB) then moved from the Open University to take up a post as Head of Mathematics at Avery Hill College and determined to pursue the initiative. It was decided to make the event one of joint sponsorship between the College and GAMMA (Girls and Maths. Association) [1] partly because this opened the network of possible contributors to the conference and partly because it would bring GAMMA to the notice of many people who might be interested in its activities.

The impetus sprang originally from a combination of factors. First, there were research studies which were suggesting that adolescent girls were more comfortable learning in single-sex classes. When offered a choice, they preferred a collaborative atmosphere in which they could explore ideas without fear of ridicule, put-down, or even simply the pressure of their social relationships with the boys.

In addition, increasing attention has been paid recently to the disproportionate rate of achievement in mathematics between boys and girls in the later years of the secondary school but, more importantly, to the increasing negativism of girls' attitudes to mathematics as demonstrated, for example, in the APU findings. Finally, girls appear to be unaware of the role of mathematics as a critical filter to careers. No longer does an 'O' level in mathematics open the way only to jobs in science and technology. It is also a requirement for jobs in, amongst others, advertising, health-service administration, personnel work, travel, and, of course, teaching, all of which might be considered areas favoured by women.

Concern with an investigational approach to the teaching of mathematics was finally also given the Cockcroft seal of approval in *Mathematics Counts,* in 1982. Investigations and problem-solving were itemised as two of the six necessary teaching styles to ensure a full range of mathematical experience. Interestingly, there has been no research which has looked specifically at gender differences either in performance or in attitude to mathematics explored in a wider way. However, those interested in this field report that girls and boys do appear to participate on a much more equal basis when they are in control of their own learning. And girls themselves are very positive about such experiences. It seemed, therefore, that to run a day-conference in which girls would be given the opportunity of meeting mathematics from a different perspective to

that normally found in schools and in which the mode of working would support and encourage them might help to start a re-perception of the subject and the individual's relationship to it. Putting this together with some information on careers seemed like a good way of motivating girls to re-think their position with respect to mathematics and, perhaps, to its relevance to them.

The day was organised, therefore, from a wide range of concerns. We were concerned that girls achieve less and feel worse about mathematics as a subject of study than boys. We were concerned that, in a time of rising unemployment, failure to achieve an 'O' level in mathematics puts girls at a distinct disadvantage in the jobs' market. We were concerned that mathematics as encountered at school gives girls, and boys, a skewed view of the subject and their relationship to it and, finally, we were concerned that the absence of a problem-solving/investigational approach might disadvantage girls, as well as boys. In the climate of an active initiative on equal opportunities in the ILEA, we were confident that schools, and pupils, would be interested in ways of implementing a more egalitarian approach to mathematics education. In the event, we were nearly killed in the rush!

Organisation

The 'Be a Sumbody Conference' jointly hosted by Avery Hill College and GAMMA took place in March this year. The audience was girls aged 12-14 — it was felt that these are the years when the decision to disengage from mathematics is made. A first 'flyer' saying little more than this was sent out to secondary schools in the local authorities near the College asking teachers to telephone and book places. Within two days we had been persuaded to raise the ceiling from 150 to 220 girls. The disappointed schools were then offered one and then a second date in July.

The format of the day was determined by our two main aims: to convey a girl-friendly message about the nature of mathematical activity, and to re-inforce the narrowing effect on career prospects of disengaging from school mathematics.

The programme offered plenary sessions to open and close the conference; but most of the girls' time was spent in mathematics workshops or at a Careers Fair.

We sent participants details of the twelve workshops. They had the opportunity to attend two during the day and were asked to select the four that appealed most. On offer were sessions on Logo, 3-D models, probability, movement, calculators, games, puzzles and the maths of fashion and fabric. Amongst the leaders were three men.

This choice brought organisational problems. As the completed choice forms arrived we realised that too many girls were opting for the same few workshops. It seemed that an accurate description of a 'worthy' session was not enough to inspire. We could have filled the fashion workshop or the one on Logo four times whilst others offering investigations were not so popular. It seemed that the selling points were references to fashion, working with a friend and anything to do with micro-computers.

Eventually all the girls were allocated to two workshops each in such a way that everyone had a first or second choice.

All girls would also come to the Careers Fair to meet women doing non-traditional jobs. Their careers would not be identified as mathematical, but did require an 'O' level. They included an electronic engineer, a furniture designer, a British Telecom engineer, a quantity surveyor, a laboratory technician, and the head of TV cameras at the BBC (a male, looking for trainees).

During the Fair each woman sat under a large label saying 'I am a ...' and the girls were free to walk around the Hall and talk to them informally.

A camera team from Chelsea College made a video of the activities which is to be used to support further school-based initiatives.

It was clear from the list of participants that the day would widen horizons irrespective of our input because the schools ranged from selective suburban girls' schools to inner city comprehensives. Some teachers had brought a whole mixed-ability class, others a 'set' or a self-selected group.

There was some disappointment amongst those who had not been allocated to both their first and second choices of workshops; but, despite this, almost all the girls reported at the end of the day that they had enjoyed what they had done.

Reactions

Asked to complete the sentence beginning 'The best thing about today was ...', girls wrote: [2]
— learning about triangles. We learnt a lot and it was fun. We made shapes and talked about them.
— workshop 9 (Imagination, Movement & Maths)

because everybody had the opportunity to get involved and it wasn't formal.

– my first workshop ... which was to do with shapes and we made a hexaflexagon and we tied ourselves up and we then had to try to get free.

– making models in workshop 5. The teachers were also very nice.

– the two workshops I went to. I was surprised how easy designing is with Maths.

– the workshops. They have helped me learn that there is more to Maths than sums.

– everything I done was good and interesting I much liked the logo I thought it was brilliant I would like to come again.

– computers, knitting and that's all.

The response to the Careers Fair was mixed. The organisation relied upon the girls being sufficiently self-motivated and self-reliant to move around and ask questions. For a large proportion of them this was outside their experience and they absented themselves, later reporting the session as boring and badly organised. For others it was the best part of the day:

– meeting women doing men's jobs and vice-versa was very good and enjoyable.

– the best thing about today was going round talking to women whose jobs we may not think of doing.

To our surprise, however, the conversations that the girls held with the women rarely focused on mathematics. They wanted to talk about the practical problems of working. 'What do they do with their children all day?' 'How do they manage the housework?'

One of the women described swift exits whenever she steered the conversation towards the importance of Mathematics to her job. Another, however, had the opposite experience.

A powerful piece of information was an OHP slide listing the careers for which 'O'-level mathematics is an entry requirement.

It is clear that the Careers Fair did need more structure and that several organisational changes must be made for the re-runs but the overall evaluation was enthusiastic, typified by:

– Today was beneficial and we're surprised how much Maths has to do with everyday life.

– I enjoyed it the people were really nice and I would love to come again.

– It was a different experience...

We had not thought that the venue was very important but for many girls it was clearly an unusual experience to be in a pleasant environment. They enjoyed walking in the grounds, winter gardens and the library between sessions. They were also trusted to find their way around a large college and to be in the right place at the right time.

It was an opportunity to meet girls of the same age but from very different schools. They were far less intimidated by this than might be expected and many reported 'making new friends' as one of the best things about the day.

They had been asked to make choices about the sort of mathematics they wanted to do. Opting-in (even if it wasn't a first choice!) was important to the atmosphere of the workshops.

The workshops were run by people without the teacher label. Several girls remarked that it felt so much better to know and use the first name of the people they worked with.

– The staff made a nice atmosphere

Their teachers weren't there! This was a contentious point with the teachers. Each school party was accompanied by at least one member of staff and a separate programme was arranged for them. This annoyed some of the teachers (and girls), who would have liked to see the activities but we believe that their presence would have inhibited the girls.

The feeling in the workshops was very relaxed. Each leader used her or his own mechanisms to achieve this but all shared the determination that everyone should be active, involved and unthreatened.

The girls were responsible for the discipline of the group. It would be gratifying to report that there were no behavioural problems. Recorded on video-tape is the evidence to the contrary. To quote:

– there are some very selfish people about who pull out the computers when others want to use them and lose other people's information.

The incident was resolved with virtually no action from the workshop leader!

Mathematics was redefined. The most striking examples of this come from the workshop on Fashion and Fabric. Here girls worked on complex problems of sleeve-fitting, fabric-matching and pattern-cutting and were confronted with the reality that they had been working on mathematics.

Mathematics was assumed to be worthy of protracted activity. One group of girls said they had not thought they could work for a whole day on mathematics, but now they wished there had been more time available. Many asked for four workshops not two or said that each session had been too short.

The 'problem' was brought out of the closet! Some girls said that Maths was not a problem for them because they were at a girls' school or because they 'did SMILE' [3]

Some girls were free from male-dominated classrooms for the first time. This was particularly striking in workshops using microcomputers where there were vivid descriptions of boys denying access to the girls.

— The best thing about today was that there were no boys and no-one got into trouble.

Girl-friendly mathematics was seen as possible. As one girl wrote three days later:

— I don't know what you did with her but Miss X has become much nicer.

Reflections

It is our firm conviction that sex-differentiation in mathematics education reflects the distribution of power in our society. The school and the classroom, and indeed, very often the family are part of, and mirror, that distribution. A female prime minister is no evidence of a re-distribution of political power. One only has to examine the numbers and roles of women in the Cabinet, in the Houses of Commons and Lords, in the Civil Service, in the top echelons of industry and, to come nearer to home, in headships and senior management in schools, to be convinced that power and decision-making still rest with men. In ILEA secondary schools, 49.7% of teachers are full-time women of whom approximately 75% are on scales 1 and 2. In primary schools, 80% of teachers are women, of whom 90% are scale 1. Only 54% of

headteachers in these schools are women. And ILEA is a better Authority in this respect than most. **Examine the sex distribution of authority in your school/governing body/inspectorate/LEA administrators. [4]**

Under such circumstances, male styles of working, male preferences and male-identified high-priority subjects are bound to achieve higher status and a greater propensity to operate. However, the economic facts are that many families are currently dependent upon women's earnings. The 1974 Family Expenditure Survey showed that without the contribution of women's earnings, the number of families living below supplementary benefit level would have trebled (quoted in [10]) and times have surely worsened since then. In these circumstances, we, as teachers, have a great responsibility to our girl pupils to ensure that, at the least, they understand and are aware of the factors which can operate to their advantage or disadvantage. **You might initiate discussions with your careers colleagues** to ensure that they are aware of the critical filter-role of mathematics and that this awareness is passed to the pupils. Simultaneously, you might **look at currently used, successful techniques for raising awareness** of the issues and changing beliefs held by parents, colleagues and pupils. You could, for example, **focus on the stereotyping of mathematics as a male domain** by showing a videotape of the 'Horizon' programme, Mathematical Mystery Tour, and concentrating on the roles women play in it and the hidden messages of the video about women's contribution to mathematics. You could **research information about women mathematicians** (see, for example, an excellent pack developed for schools in Victoria, Australia, on Women in Maths and Science) and point out how their contribution has disappeared from the history of mathematics. **Examine the textbooks and resources** you use for sex-stereotyping. This is particularly apparent in the illustrations of Primary schemes and the impressions given by distributors of resources that active, scientific and enquiry-based equipment is most appropriate to boys. **Investigate the feelings of your pupils towards mathematics** by putting them into groups and asking half the groups to discuss and report on the subject of the following anecdote:

'John won a prize for mathematics last term. Describe John !'

Ask the other groups to do the same but with Anne as the subject.

Attitudes towards mathematics can be tested by attitude questionnaires which can then be discussed openly in the classroom. Evidence suggests that a fear of success operates amongst girl mathematicians depressing their mathematical performance and often leading them to deny interest or ability in an area which is conventionally identified as male.[9] In addition, there is substantial evidence that boys and girls react differently to success and failure, boys attributing success in mathematics to ability, girls to effort, boys attributing failure to lack of effort or bad luck, and girls to lack of ability. It is noticeable that girls' success is attributable by them to external factors whereas their failure is internalised. For boys this pattern is reversed. Teachers who are aware of these dangers can **try to avoid reinforcing the attribution patterns** by making factual statements and avoiding judgmental ones.

When girls are given the space to listen and exchange with one another in a collaborative way they often change their attitudes to mathematics. When a boy is a member of the group, there is a tendency for him to hijack control of the group and dominate it. This is consistent with the evidence that teachers' responses to boys are significantly more numerous than those to girls and that this is partly due to the demanding behaviour of certain boys who dominate the interactional exchanges. This is observed whatever the age of the children, from the nursery upwards. **Look at the use of the computer by girls and boys.** Who is controlling the action in a mixed-sex group? If the calculators are handed out, who uses them? Changing the pattern of interactions from class or individually-based to group-based removes the possibility of class dominance from an individual pupil and places pupils together into a situation where they need to develop the skills of working together, listening and evaluating what each member of the group says or does. These skills, in our society, are regarded as female. 'Those who have their language, their experience and themselves rejected ... seek protection in silence. Those who have their experience, their language and themselves validated ... are encouraged to assert themselves even more' (Dale Spender in [10]). In single-sex classrooms, these differences are still noticeable and just as reflective of different experience and different feelings about the subject-matter. **Who is silenced and who is validated in your classroom?** Ask a colleague to observe.

There are many ways in which sensitivity to sexism in the mathematics classroom can be developed. We have only referred to a few here but an in-service pack is in preparation to support teachers who want to focus on these issues. We would only want to conclude by saying that it is our experience that once you begin to notice sexism, you cannot stop and the will to do something about it is generated. In the process of removing or challenging power, some people are going to feel their loss, and resist. That is to be expected and strategies must be developed for dealing with such resistance. We append some references which we have found helpful in establishing some facts in place of myths. But most of all, we recommend that you talk to your girl pupils and hear how they think it is and work together with them to change it. ∎

Avery Hill College, London

Notes

1 To join GAMMA contact Marion Kimberley, Goldsmiths College, London
2 The quotes are from the evaluation forms which the girls completed at the end of the day.
3 Secondary Mathematics Individualised Learning Experiment developed in ILEA.
4 In this section we adopt the 'Cockcroft convention' and indicate with bold print suggestions for action.

Further Reading

5 Brush Lorelei: *Encouraging Girls in Mathematics: the problem and the solution,* Abt Books, 1980
6 Deem Rosemary: *Schooling for Women's Work,* Routledge and Kegan Paul, 1980
7 Delamont Sara: *Sex Roles and the School,* Methuen, 1980
8 Eddowes Muriel: *Humble Pie,* Longman for Schools Council, 1983
9 Horner M.S.: *Towards an understanding of achievement-related conflicts in women,* Jour. of social issues, 1972, 28, 157-75.
10 Marland Michael: *Sex Differentiation and Schooling,* Heinemann, 1983
11 Walkerdine Valerie and Walden Rosie: *Girls and Mathematics: The Early Years,* Bedford Way Papers, 1982
12 Weiner Gaby: *Just a Bunch of Girls,* Open University Press, 1985

Script by Diane Finch

THE LEARNING SUPPORT
OF VERY ABLE PUPILS

Although this is an account of how we have tried to develop a policy for the learning support of very able pupils, it is important to point out that this was not the original intention.

As head of maths I have been working with the members of the department to produce a variety of policies, some of which are addressed to particular groups of pupils. Initially there was a need to ensure a clear purpose in the learning support for the least able, particularly as this was a shared responsibility with our learning support department. This was followed by extensive work on a policy to redress the gender bias in our external examination results, which took the best part of a year to achieve.

A major initiative within the school at this time was the establishment of a multicultural working party, and the experience of working with this group broadened the outlook of all involved. Within the mathematics department, we began to encompass all of the previous work on multicultural, gender and, learning support into an Equal Opportunities statement.

It was at this point that I became aware (at last!) of the fact that we have a small but growing number of highly able pupils in mathematics. These pupils have difficulties particularly because they are very able relative to their peers and are capable of offering a variety of challenges to their teachers. It has been helpful to identify some of the challenges.

- how do we keep these able pupils working to their full potential
- can we face up to, and live with the fact that some pupils may be more able (within their limited knowledge) than their teacher
- how do we protect these pupils from inappropriate peer group pressure
- how do we encourage staff to avoid complacency, since many of these pupils are obviously, almost by definition, getting on alright anyway?

I raise these points to warn of some of the dangers. Within our mathematics department we have worked as a consistent team with few personnel changes over a number of years. I felt that we discussed these points in an open way, without anyone feeling unduly defensive. This may in part be due to our previous experiences of working on other equal opportunities policies, and in part because the work on very able pupils was placed in the context of equal opportunitles.

After several meetings we arrived at the policy which is outlined below. It is more important to go through the process than to arrive at a policy, so please take this simply as an example.

Policy for learning support of very able pupils

Why do we have a policy?
- because we have a proportion of very able pupils who go on to take mathematics and maths-related subjects at university
- because we recognise that learning support is required for the extremes of our ability range, both in order to give the fullest range of opportunity and as part of effective classroommanagement
- because it is important for staff, pupils and parents to recognise the high levels of attainment that are possible and that are being achieved
- to encourage excellence in mathematics and to promote the future development of the subject
- because of our desire to assist every pupil achieve and extend their personal best
- because we need to develop means to support the most able pupils from occasional inappropriate peer group pressure.

How do we identify very able pupils?
We don't! First we assume that we *do* have very able pupils, and then we set work using the neutral stimulus approach, i.e. work which is accessible to all pupils. For example ... how many squares on a chessboard, ... how many different games of noughts and crosses are there . . . ? Very able pupils *identify themselves* under such circumstances.

Forms of learning support that we want to offer are as follows;
- the continued identification of suitable extension work for the main school topics
- extra topics to be covered in class time while others are working on the main topics

- there will be an occasional maths club run on an invitation basis
- where classes are simultaneously timetabled, the occasional grouping of classes should be used to permit extraction of pupils as well as a variety of other benefits
- the continued use of the neutral stimulus approach
- the most able pupils should of course, often not need to cover the easier work, and can be started further up the difficulty slope
- including specific provision in teachers' guides
- we could consider the development of personal projects which could form the basis of magnificent pieces of GCSE course work
- take the exam early, although there are practical difficulties with this which necessitate caution
- we will review the possibility of able pupils taking the new extension paper avaiable from MEG as an extension to the GCSE course
- the development of positive peer groupings to provide mutual support. This is difficult and potentially dangerous, but it is essential to counter some of the negative atttitudes that a handful of pupils have to those that do work hard and succeed.
- encouraging a whole school policy for the support of the most able.

The first stage in trying to achieve a whole school policy was simply to ask the headteacher for approval in principle to the idea of beginning the process. The head greeted the suggestion with enthusiasm and promptly placed it on the agenda for the next heads of faculty meeting. At the time of writing we have a draft school policy awaiting further discussion.

HIDDEN MESSAGES

Jenny Maxwell

There is an almost unchallenged assumption that mathematics education, for both teacher and taught, occurs in a political vacuum. This I cannot accept: it seems impossible that such a central part, mathematics, of such a political institution, education, should really be politically neutral.

It is easier to be objective about, and therefore to recognise, the social bias of mathematics questions from abroad. Chinese examples stress agricultural and military applications; a Cuban textbook asks children to find 'the average monthly number of violations of Cuban air-space by North American airplanes'; Russian children are asked about collective farms; East German children have a similarity problem about a tower in Berlin and a bigger one in Moscow. All the following problems contain assumptions, some explicit and some implicit, about the society from which they (or in some cases their enemies) come.

> Twenty-three peasants are working in a field. At midday six guerilla fighters arrive to help them from a military base near to their village. How many people are working in the field.
>
> (Mozambique)

Once upon a time a ship was caught in a storm. In order to save it and its crew the captain decided that half of the passengers would have to be thrown overboard. There were fifteen Christians and fifteen Turks aboard the ship and the Captain was a Christian. He announced that he would count the passengers and that every ninth one would be thrown overboard. How could the passengers be placed in the circle so that all the Turks would be thrown overboard and all the Christians saved? (USA)

A Freedom Fighter fires a bullet to an enemy group consisting of twelve soldiers and three civilians all equally exposed to the bullet. Assuming one person is hit by the bullet find the probability that the person is a) a soldier, and b) a civilian? (Tanzania)

When worker Tung was six years old his family was poverty-stricken and starving. They were compelled to borrow five dou of maize from a landlord. The wolfish landlord used this chance to demand the usurious compound interest of 50% for three years. Please calculate how much grain the landlord demanded from the Tung family at the end of the third year. (China)

These examples coming from foreign cultures strike most of us as blatantly political, a part of the indoctrination of the young into the currently dominant values in these societies. But I feel that we are less aware of this same process when it occurs in British schools.

To find out more about this I recently made up a small collection of questions on percentages. They were based on textbook questions, some almost as printed, but slightly altered to make political points. I then asked some twenty-five teachers (in schools and FE) for their reactions to these questions.

One question concerned Mr Jones who owned a factory employing 100 people. He drew a salary of £15000 pa, paid each of the 20 supervisory staff £10000 pa and the other 80 employees £6000 pa. The question asked for the total wage bill and went on:

> The company has done well in the last year so Mr Jones decides to give himself a 15% rise, the supervisory staff a 10% rise and the non-supervisory staff a 5% rise.

There were then questions about the new wages bill. Only ten out of my twenty-five teachers found the social and political assumptions here worth commenting on:
− There's a bit to talk about there ... we might have a little talk about industrial justice
− This is the shocking one ... I'd have to have a laugh about it ..... I couldn't resist a comment.

It was surprising that there was divided opinion, even amongst teachers from the same institution, about whether or not the students would notice the inequalities of this industrial situation.

Another question was ridiculous, about a spider's weight which increased to 500 gm. Sixteen teachers commented on this absurd situation, six more than had mentioned the social context of the previous question. Was this because the first consolidated their own view of the world?

One question was about a man who won £2000 in a competition and the way in which he shared the prize money, his wife getting nothing and each of his sons more than their (older) sister. No one mentioned the latter point and only four teachers the former. Again, did they not notice or was it that this is how they know things are? One teacher made a different point: I don't believe in competition ... I usually say so.

It was alarming that only about half the teachers responded adversely to the use of the word 'alien' in this question:

> Assuming that the number of aliens in the UK is ⅓% of the population and that a football crowd is a random sample of the population, how many aliens would you expect to find in a crowd of 60000?

– I would avoid [it] .... [It is] likely to cause embarrassment to certain people and give rise to the nastier feelings of one or two members of our society.
– Calling people 'aliens' smacks of racialism.
– We'd be on very dangerous ground .. to use words like 'alien' or even to draw to children's attention that our society is a racial mix. We've got quite a few pupils in our school who will seize on anything like this as a means of causing friction between the various groups.

But what of the half who did not comment? They came from a variety of schools and backgrounds and it is difficult to believe that their pupils are more immune to racial prejudice than those mentioned above, or that such wording does not encourage prejudice, albeit subconsciously. Indeed on two occasions a teacher accepted the question without comment while another in the same school mentioned the extreme dangers of its use.

Just under half the teachers I spoke to broadly agreed with the two who said of this collection of questions: 'very establishment' and 'obviously class-biased, sex-biased and race-biased'. But the rest either did not find them so, or did not consider it relevant to their teaching of mathematics. It would have been interesting to see the teachers' reactions to questions about profits from burglary or tax evasion rather than investment. Even more revealing, had I been asking the questions now, would have been the reaction to this one, suggested by Griffiths and Howson [3].

> A coal-mine employs 1000 men and loses £N per year. If it were closed down suppose 750 of the men would not expect to find another job and would have to live off social security payments. What value of N makes it cheaper to the state to keep the mine open? Is this a good way of thinking about the problem?

Would *you* use this question? Why? Or why not?

Swetz [1] found a problem in a Tanzanian textbook about canned peaches and 'is concerned about asking a poor man to struggle through the problems of the rich'. Likewise British textbooks use questions about mortgages, investment and interest for those whose families, or who themselves, are on social security. In fact the time has perhaps already come when some people would find questions about wages offensive. Few questions ask the rich to struggle through the problems of the poor.

After further discussions with the teachers covering attitudes, methods and topics ('If a few more people had understood what inflation meant, they might not have won the election') about a quarter of the teachers remained clearly of the view that mathematics education is, and should be, politically neutral. It was interesting to see the pervasive and unanimous attitude of guilt and apology whenever a teacher felt she was questioning the norms of society. Four teachers expressed fear of being thought 'leftist'. Yet none was anxious about upholding the values of the right.

For some the discussions, and particularly the examples, caused a shift of position:

– Just looking at these questions ... one can ... have an influence even through mathematics which I see as being unlike many other subjects ... You've exposed to me that it is quite easy to subconsciously ... accidentally, inadvertently ... put forward views which .. you may not believe in. But .... through a degree of thoughtlessness and ill-considered preparation you may end up putting forward social views that you disagree with.

But no one went as far as the teacher quoted by Len Masterman [4]

> For over twenty years I worked under the delusion that I was teaching maths. The social pressures I put upon the kids were designed to make my maths teaching more effective. I now realise that I was really teaching social passivity and conformity, academic snobbery and the naturalness of good healthy competition, and that I was using maths as an instrument for achieving these things.

For me, the final proof that mathematics education is by no means neutral came in answer to the question 'Has mathematics a role to play in furthering social causes and political understanding?'

Two teachers, from the same institution, replied: 'Yes definitely' and 'Oh, I shouldn't think so'.

How can anything which can elicit two such opposing but adamant replies be neutral? Politics is about conflict and there was conflict here. ∎

References
1 Frank J Swetz: *Socialist Mathematic Education*, Burgundy Press 1978
2 Paulius Gerdes: *Changing Mathematics Education in Mozambique, Educational Studies in Mathematics*, vol.12 1981.
3 H B Griffiths and A B Howson: *Mathematics, Society and Curricula*, CUP 1974
4 Len Masterman: *Teaching About Television*, MacMillan 1980

EDUCATION FOR EQUALITY AND JUSTICE: PART ONE
Whose Mathematics? – What Mathematics?

Sharan Jeet Shan & Peter Bailey

We begin our three part discourse with an overview of some general attitudes to the culture of mathematics and mathematics teaching as well as the cultural context in which mathematics develops and grows.

Mathematics, one of the three core subjects in the NC is considered by many to be a culturally neutral area of study. The *concise Oxford dictionary* defines it as a 'pure, abstract science of space and quantity'. Yet no other subject inspires so much debate, controversy and research at all levels amongst those who are involved in its teaching and learning. In no other area of study do so many achieve so little and judge their overall success in school by success in mathematics. This sad fact has inspired many maths educators to examine the context of mathematics so that the quality of learning can be improved by making it closer to the real lives of their students, thereby improving examination results. At present, the world over, there is an explosion of this debate amongst mathematicians around the cultural influences on mathematics as a body of knowledge and its implication for classroom mathematics – Hudson, Bishop, Gerdes, D'Umbrasio, Frankenstein – to name but a few.

There are two main historical reasons why this daily life activity has come to be incorrectly regarded as 'pure and neutral', developed mainly by the West and mainly by men.

Firstly the study of mathematics has been accorded a mystique and a high learned value, mostly from high class/high caste men who communicated their discoveries to each other under oath. Hindu and Arab mathematicians saw mathematics as holding the key not only to the stars and planets, but also to the timeless truths of gods, heaven and hell. Position of the male in ancient Hindu, Arab and later Greek and Roman societies meant that mathematical problem-solving, was and even today is, confined to traditionally 'male activities'.

Secondly, the European dominance of the past 400 years suppressed high achievements in mathematics and science in colonised societies thus embedding mathematics into a eurocentric mould.

In addition to this, the fundamental problem for teachers in our view is that the notion 'teachers of children' is replaced in those who teach mathematics by a deeply held belief that they are teachers of mathematics – an abstract, universally-known body of knowledge. The question 'Why did I become a teacher?' is rarely asked. 'Why do I continue to pursue such a low paid, under-resourced, undervalued profession?' is asked sometimes. In many workshops that we have done, maths teachers are often surprised and at first puzzled at being asked such a question. These questions must be asked for there is no other profession in which this generation holds such a privileged position over the next. Often pupils who are 'street-wise' and can do very complicated exchange, conversions, calculations at a market stall are unable to put their maths into school form. What does this say to them about school and school mathematics in particular? Our students' values, attitudes, beliefs and future careers are largely decided by the knowledge that we impart to them in a closely knit teacher - pupil relationship within the narrow confines of a classroom. This interpersonal exchange is little influenced by educational changes at a political level. We alone have the responsibility for that. In Sanskrit the word 'teacher' means 'Guru' and a very high position is accorded to a Guru. A Guru has the responsibility of engaging the 'Shishu' (meaning pupil) in a critical dialogue helping the pupil to develop his/her faculties to their maximum. The learning is not complete until the 'shishu' can challenge the 'Guru', though showing extreme respect and reverence.

We would therefore like you to find a quiet half hour and write down 10 reasons to both the two questions:

1. Whose mathematics? What mathematics?

2. How many of the reasons why you became a teacher manifest themselves in the reality of your mathematics classroom?

The cultural context within which mathematics developed and is taught also requires thorough examination. Social scientists have many diametrically opposed definitions of culture but we do not wish to enter that debate. Culture is a living, growing dynamic phenomenon. Thousands of our inner city pupils have helped us to see that a fossilised preserving notion of culture excludes them. Of course they are as much 'British' as pupils living and learning in predominantly white shire counties. In every human being her/his culture is the embodiment of past and present, home and school, parental and indigenous culture, collective family as well as personal experience, thus making it a unique entity. Around Britain there is a wide variety of 'traditional' culture such as rural, northern, Welsh and Cockney. In communities that have migrated in large numbers such as Kenyan Asians, Chinese Jamaicans, Black Americans, Scottish Muslims one can demonstrably see mixed loyalties to their multilingual, multicultural background. It is this understanding which has changed our formalised way of teaching such mathematical topics as ciphering, counting, classifying, ordering, inferring and modelling. Mathematics is used freely as a means of communicating many aspects of our collective history as well as present trade and aid, news and views.

Before we can make a start at 'multiculturalising' our classroom mathematics in accordance with the needs of our pupils, the question that needs to be asked is: 'What is *my* culture?' Here is an example written by Sharan-Jeet.

> I am a Punjabi Indian by birth, born into a very middle class, authoritarian family. My father was highly educated. Learning of science and mathematics dominated our household. I was the eldest child. At school and home there were strictly defined roles for females and males. I did not come to this country by choice but had to follow my husband. I have two sons. I have been a single parent for the last 15 years and I have lived and worked more than half my life in England. I am a professional educator and I take a keen interest in politics. *This is my culture. What's yours?*

It is only when we go through the process of understanding the total past and present culture of our pupils that we will acknowledge cultural-specific ways of learning mathematics as equally valid, interesting and motivating, both for our pupils and ourselves. Only then can we move away from the narrow mould of Rangoli patterns and Islamic patterns. Mathematics used by women, farmers, fishermen, weavers, gamblers, street traders is a panorama full of surprises, intrigues and fun for all our pupils. We have explored some of these in our book: *Multiple factors, Classroom mathematics for equality and justice* (Dec.90).

Finally we would like you to explore another dimension to the change process. Ask the question 'Why do I want to change my approach to teaching mathematics?' You have to decide if you simply wish to 'multiculturalise' your mathematics, or to use mathematics teaching for raising issues of equality and justice. For this, classroom mathematics, as one part of the whole curriculum, needs not only to be set into the cultural but also the social, political and economic context of contemporary society. Our students often appreciate the role of mathematics in society but rarely do they gain confidence to use mathematics to enhance their own role.

In order to examine one's current approach, and decide what measures to take, questions need to be asked.

– Which pupils benefit most in present-day teaching and examining?

– How do they benefit?

– Are the pupils set? What are the criteria for setting? Are these fair?

– Are there any aspects of my classroom which strike 'fear' and 'dislike' of mathematics in my pupils?

– What is the gender balance in my pupils? In textbooks?

– How are issues of race presented?

– What is the real received view by students of mathematics?

– Does my classroom mathematics relate to the world of trade and aid?

– Does it relate to technological developments?

– Does it reflect the 'reality' of local, national as well as wider world issues?

– Does it challenge as well as motivate?

It is possible to look at 'culture' without challenging. We believe that such an approach can do more harm than good as it has the danger of being tokenistic, patronising and condescending. Such an approach should be avoided. ∎

Sharan Jeet Shan
General Adviser – Assessment
Peter Bailey
St. Albans, Birmingham

EDUCATION FOR EQUALITY AND JUSTICE: PART TWO

Appropriate learning context for raising issues of culture equality and jústice

Sharan Jeet Shan and Peter Bailey

This article looks briefly at the influence of social and cultural relations on the *process* as well as the elimination of bias in the *content* of mathematics lessons.

Process

The process of delivery is considered at two levels, the learning that takes place in the presence of the teacher and the definitions, and reflection in classroom images. These are crucial factors in deciding upon students' perceptions of mathematics. Both should be largely decided by the culture and personality of the pupils and not simply by some prescribed syllabus or National Curriculum requirements. Experience shows that teachers directly influence their pupils' like or dislike of mathematics. *Better Mathematics: A Curriculum Development Study* emphasises personal involvement by the teacher as an important factor in motivating and encouraging. By getting to know and including the culture of our pupils in their learning, we can spark their interest more effectively. Aspects of teacher-pupil interaction which can be isolated as being strongly influenced by culture are:

– language in the mathematics classroom;

– teachers' perception of 'race', culture and achievement;

– classroom techniques.

Language in the mathematics classroom

English is the medium of instruction for teaching mathematics in many countries. In England we notice one vital difference. While in many countries being bilingual in, for example, English/French or English/Punjabi or English/Chinese is considered to be advantageous, in England it is generally perceived to be a problem, particularly if the bilingual speaks non-European languages. Whatever the historical facts, English people seem to have a general dislike for learning other languages. This sad fact has led to problems for many inner city children who often come from multilingual backgrounds and find that their bilingual achievements are not recognised by teachers. They end up rejecting their home langauge as a useful tool for learning. In many cases they end up in low maths sets because the teacher only considers the level of English and takes no account of other language skills. Of course, many teachers have rejected local dialects as a vehicle for educational learning, relying on 'standard' English alone. As a result thousands of working class children too have underachieved because they have been taught by teachers who have low expectations of them.

Language and culture may influence logic and reasoning system, so often fundamental in mathematics. The Cockcroft Report (DES 1982) para 245 and 306 - 310) asks teachers to be aware of:

– variation in the level of language skills of students;

– the variety of language used for a particular mathematical expression;

– some stylized expressions which may make it difficult for students to comprehend the problem and, therefore, choose the correct solution.

To this list we would add this other factor concerning students who have English as their second language. Lack of understanding in English must never be mistaken for poor conceptual development. It is possible that students base their interpretation of new words on words they already know. An example from science illustrates this for E2L students. *Water* in some languages is used for *liquid* in some circumstances.

Strevens (1974) listed the following as key issues for mathematics teaching in Anglophone developing countries. Teachers in all schools need to consider:

1 Do the teacher and learner share the same (first) language?

2 Do the teacher and learner share the same culture?

3 Do the teacher and learner share the same logic and reasoning system? (And is this the logic and reasoning system we find reflected in mathematics?)

4 Is there a 'match' between the language, culture and logic/reasoning system of pupil and teacher?

Oral exchanges between students who speak the same first language assist in remedying this situation. The biggest advantage with developing Asian bilinguals is that many English words have been absorbed into everyday vocabulary. A bilingual mode of delivery would greatly benefit such pupils. For the sake of all our pupils, therefore, language should be simplified without diluting the concepts involved. An uncluttered layout, interesting pictorial presentation, familiar ideas, active use of all languages in the classroom – are all factors which will provide support for the learner. The SMILE scheme is a good example of this. Approaches such as these benefit all students with restricted vocabulary.

'Race' and achievement

The second factor which influences teacher delivery is teacher's perception of 'race' and 'achievement'. Much has been written about the manifestations of racial inequality in schools; see for example (Klein (1985) Arora and Duncan (1986) Eggleston (1986) Gaine (1987)). Very little is written about the effect of racism on students both black and white, even less about how positive perceptions and delivery will enhance the motivation and hence achievement of underachieving pupils in mathematics. There is some research available on gender inequality. We can explore the parallels and learn from this. Besides, such influences can only be understood when put into the *whole* context. We have also found that misconceptions about 'race' and 'achievement' surface sharply when we first introduce the notions of equality and justice to maths teachers. Racism, sexism and any other inequality has to be challenged as much in a maths classroom as in any other. If the teacher's delivery is such that positive images are presented of all groups in society, it can only assist in reducing the friction present in a classroom.

Classroom techniques

An article by Derek Woodrow in *Mathematics in Schools* best illustrates the third factor: teaching styles and classroom techniques.

> As it is currently taught, mathematics in schools demands concentration, self discipline, accuracy, conforming to rules, quietness, precise and

sophisticated language. Why is not mathematics taught so as to encourage creativity, group cohesion, intuition, expressiveness and extroversion?

Traditional-style maths lessons presented five times a week for eleven years is for many a sure way to turn pupils off maths for ever. The variety asked for by the Cockcroft report should also be evident in social and cultural contexts.

Content

In looking at the content of maths textbooks and lessons, one must acknowledge the fact that through this media children receive very powerful messages of race, class and gender classifications. It is important to check carefully to see what is included, what is excluded, what is misrepresented and why it is useful to redress the balance.

Here is a brief checklist of what textbooks and other influential materials of the classroom (exam papers, wall charts for example) should include (see also Hudson (1987));

– pictures and settings depicting diverse positive images of people of many ethnic groups and social classes experiencing a variety of life style, food, work, hobbies and preferences;

– mathematics applied to many different areas of life such as science, sport, industry, politics and religion;

– using up to date statistics to challenge racism and sexism;

– acknowledgment of social inequality: rich and poor, young and old, healthy and poor living standards and their causes, starvation and greed;

– acknowledgement of the contribution to mathematics of a variety of past and present cultures to mathematics and to ways of solving problems. ∎

Sharan Jeet Shan
General Adviser, Sandwell LEA

Peter Bailey
Deputy Head St. Albans School, Birmingham

References

R Arora and C Duncan (1986) *Multicultural education: towards good practice*, Routledge and Kegan Paul, London
DES (1982) *Mathematics counts* (The Cockcroft Report) HMSO, London
B Hudson (1987) 'Multicultural mathematics' in *Mathematics in schools*, 16 (4)
P Strevens (1974) 'What is linguistics and how it may help the Mathematics teacher?' UNESCO Paris. An introductory paper prepared for the 1974 Nairobi Conference
D Woodrow (1974) 'Cultural impacts on children learning mathematics' in *Mathematics in schools* 13 (5)

CLASSROOM MATHEMATICS FOR EQUALITY AND JUSTICE

Sharan-Jeet Shan & Peter Bailey

In the previous two articles we attempted to clarify the kind of learning context required in a mathematics lesson if a teacher wishes to challenge the neutrality myth and eurocentrism of classroom mathematics. Only then can students be encouraged to seek their own solutions for the world

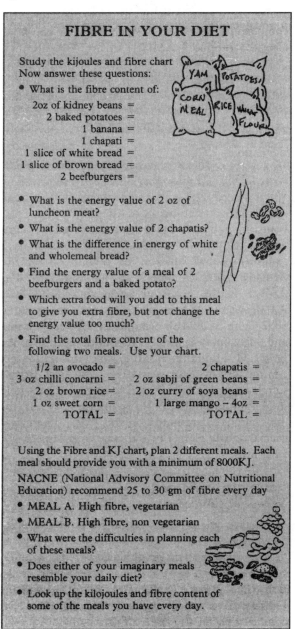

FIBRE IN YOUR DIET

Study the kijoules and fibre chart
Now answer these questions:

* What is the fibre content of:

  2oz of kidney beans =
  2 baked potatoes =
  1 banana =
  1 chapati =
  1 slice of white bread =
  1 slice of brown bread =
  2 beefburgers =

* What is the energy value of 2 oz of luncheon meat?

* What is the energy value of 2 chapatis?

* What is the difference in energy of white and wholemeal bread?

* Find the energy value of a meal of 2 beefburgers and a baked potato?

* Which extra food will you add to this meal to give you extra fibre, but not change the energy value too much?

* Find the total fibre content of the following two meals. Use your chart.

  1/2 an avocado =   2 chapatis =
  3 oz chilli concarni =   2 oz sabji of green beans =
  2 oz brown rice =   2 oz curry of soya beans =
  1 oz sweet corn =   1 large mango – 4oz =
  TOTAL =   TOTAL =

Using the Fibre and KJ chart, plan 2 different meals. Each meal should provide you with a minimum of 8000KJ.

NACNE (National Advisory Committee on Nutritional Education) recommend 25 to 30 gm of fibre every day

* MEAL A. High fibre, vegetarian
* MEAL B. High fibre, non vegetarian
* What were the difficulties in planning each of these meals?
* Does either of your imaginary meals resemble your daily diet?
* Look up the kilojoules and fibre content of some of the meals you have every day.

KILOJOULES AND FIBRE CHART

FOOD	PORTION	KJ	FIBRE in gm
Aubergines raw	7oz (200gm)	120	5.0
Avocado	1/2 – 3 oz	860	2.0
Bacon	100 gm 1 back rasher		
	raw	600	0
	grilled	340	0
	fried	380	0
	1 steak grilled	420	0
Banana	6 oz (100gm)	320	3.5
Mango RAW 1	201 gm	152	6.0
Honeydew Melon	149 gm	49	5.0
Papya RAW	304 gm	119	4.8
Raisins Seedless	145 gm	419	1.0
Strawberries	149 gm	55	4.2
Barcelona nuts	1 oz	720	3.0
Beans			
Baked	8 oz	580	16.5
Black eyed	1 oz	380	7.0
Butter	4 oz	440	5.5
Red Kidney	1 oz	300	7.0
Runner	4 oz	80	4.0
Soya	1 oz	447	4.0
Beef:			
Beefburgers	2 oz	520	0
Corn-canned	2 oz	482	0
Fore-rib roast	1	760	0
Stewing steak	3 oz	740	0
Beef sausages	1 large grilled	520	0
Beef & Pork	1 large grilled	540	0
Mince beef	3 oz	720	0
Brazil nut	1	80	0.5
Beet root	2 oz	100	1.5
bread:Brown	2 average slices	680	3.5
white	2 average slices	640	2.0
Wholemeal	2 average slices	860	9.0
Chapatis			
Rice: Brown	2 oz	840	2.5
White	2 oz	866	1.5
Lentils scup	2 oz	500	5.0
Black-eyed peas	150 gm	190	2.5
Mung beans	105 gm	355	4.6

around them. This closing article gives a few examples which can be incorporated in a maths lesson to provoke thought. Each example can be extended into other subjects. Examples are taken from our book *Multiple factors: classroom mathematics for equality and justice* (1).

Food is a central theme in issues of poverty, famine, war and overpopulation. Asian and Afro-carribean diets are considered deficient in vitamin D and proteins, *smelly* – full of spices such as garlic: leading to comments like *Indians smell of curry Miss!*

Fibre in your diet is a cross-cultural and cross-curricular look at planning for a balanced diet.

Extension work can explore over and under-consumption or challenge myths about world food production using these two graphs.

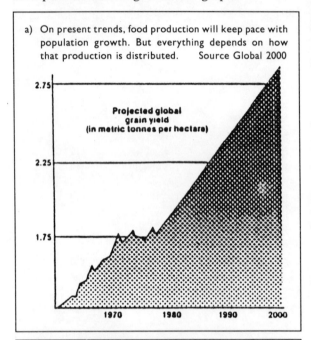

a) On present trends, food production will keep pace with population growth. But everything depends on how that production is distributed.   Source Global 2000

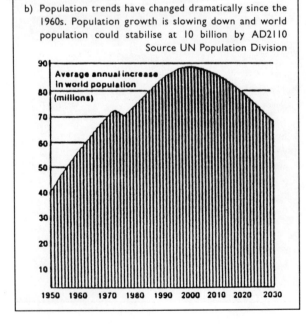

b) Population trends have changed dramatically since the 1960s. Population growth is slowing down and world population could stabilise at 10 billion by AD2110
Source UN Population Division

In *Transport graphs* the idea is to look at the reality of fuel consumption for different forms of transport. Cars and planes are a vital part of our lives. It is essential to understand the costs involved.

PICTOGRAMS AND DIAGRAMS

Q. Use a horizontal bar chart (with pictures if you wish) to illustrate this data about transport. The information shows how far one person can go on one gallon of fuel.

TYPE OF TRANSPORT	PASSENGER MILES/GALLON OF FUEL
Commuter train	100
City bus	95
Underground train	75
747 jet	22
Car (5 people – between cities)	20
Concorde	14
Helicopter	8

Now do some calculations which show how much fuel is needed and the cost incurred for each mile in the different forms of travel. Design another diagram to illustrate this information in an interesting and accurate way. (See *The new state of the world atlas* published by Pan Books for numerous ideas on information diagrams.)

Q. How do you think the following forms of travel will compare?

Bicycle	Walking
Horse	Camel
Horse and Cart	Rickshaw
Donkey	Mule
Horse-drawn canal barge	Solar car
Milk float	Motor bike
Hoverboard (as in Back to the Future II)	

Our aim is to prompt our students to ask their own questions. We feel that differing perspectives emerge naturally when the context is challenging.

We do not over-emphasise right or wrong answers nor have long imposed dicussion in a maths lesson.

We hope that our articles have made you curious and that you will want to try to use mathematics to raise issues of equality and justice in your classroom.   ■

Sharan-Jeet Shan
General Adviser (Assessment),
Sandwell Education Authority

Peter Bailey
Deputy Head, St. Albans School, Birmingham

References
S-J Shan, P Bailey *Multiple factors: classroom mathematics for equality and justice,* Trentham Books 1991

THE REBIRTH OF EDUCATIONAL SOFTWARE

Richard Noss

The last decade has seen the mathematics education community struggling valiantly to outlive the legacy of the early days of computers in schools. The mushrooming in the early eighties of count less programs thrown together by well-meaning amateurs, claiming to teach this or that mathematical topic—programs which did much to turn a generation of mathematics teachers against computers altogether—has given way to a much more concerted effort to try seriously to address the problems of mathematical learning from perspectives other than outmoded behaviourism.

The pages of *Micromath* and other journals world wide testify to the success of this effort: the mathematics education community is sharing an increasingly broad experience of children express ing themselves mathematically with computers, in ways which would be difficult or impossible with out them. One major trend, particularly in the UK, has been the turn *away* from educational software, and towards the harnessing of powerful computational tools such as spreadsheets and databases as well as Logo. Meetings of teachers and advisors now routinely think in terms of a requirement for four or five pieces of software for the mathematics classroom (primary and secondary), whereas ten years ago it was not uncommon to hear estimates in the hundreds.

One symptom of this change is that many mathe matics educators - and I include myself - have been ready to accept the commercial antecedents of such software in preference to much which professes to be educational. While it is clear that, say, a commercial spreadsheet such as *Excel* can hardly be considered sensitive to every educational need, it provides sufficient power and flexibility to allow teachers and children to use it in expressive ways, and to build exploratory settings (together with books, materials etc.) which exploit some thing like its full potential as a piece of software.

On the other hand, it remains true that commercial applications fall short of the mathematician's ideal in a number of ways. Some of these are technical: who in the educational world, for example, would have chosen the arcane and PASCAL-like syntax of *Excel* as the programming language underlying it's macros? How could we justify, say, the extraordinary difficulty of making a commercial database do anything other than routine tasks?

Other limitations are cultural. For example, the difficulties of programming (read, 'configuring') a database are a reflection of their underlying design priorities: programming a commercial database is a task for professionals who configure them for key board operators - possibly an efficient procedure in banks and building societies, but hardly ideal in the mathematics classroom.

I think that the immediate future will indicate yet more ways in which such programs can be harnessed by the mathematics teacher (the latest example is the family of computer algebra systems - Mathematica, Derive, Maple etc. - none of which were designed expressly for educational settings). But looking to the middle and longer term, I think I can discern a trend to return to software which is designed by educationalists for educational purposes .

The seeds of this (counter-) revolution have been sown in the first ten years of educational computing, and they are now beginning to germinate. The first of these has consisted of what have - until now - been called mathematical tools; the second has involved programming.

BEYOND MATHEMATICAL TOOLS

Mathematical tools have been the subject of discussion and implementation from the beginning, two obvious examples being the ubiquitous (and useful) graph plotters, and statistics programs. These programs are pre-tuned to specific mathematical tasks: they allow exploration in a specific mathematical domain, but one which may be functional in a wide range of mathematical situations based on that domain. The metaphor of tuning is designed to paper over the question of whether or not the programs are content-free or topic-based, designations which are not

111

obviously useful, and which mistakenly focus on the software rather than on the mathematics: clearly a graph plotting program is about graphs, even though graphical analysis might be useful in a range of different situations.

These kinds of programs are now spawning a new generation of software which genuinely harnesses new developments in hardware, and - crucially expertise in educational software development. Typically, they are developed over a period of years, on computers which are not in widespread school-use, by teams numbering in tens and with realistic budgets. Often, they are the product of collaborations between hardware manufacturers, programmers, and educationalists. Always, they are tried out in educational settings, by teachers and researchers.

Four examples will serve to illustrate the genre:

• *Multimap* 1 is a program which allows mappings to be iterated graphically. It can be used as a simple transformation geometry tool (functions can be iterated just once!), or as an exploratory tool for topics such as linear algebra and non-linear dynamics. *Multimap* allows the specification of functions in terms of translations, rotations and scalings, and its interface provides straightforward control of images, as well as direct manipulation devices, measuring tools etc.

• *Elastic* [2] (Environment for Learning Abstract Statistical Thinking) provides students with powerful tools for exploring objects and processes involved in statistical reasoning. It is a tool for entering, manipulating and displaying data. Histograms, barcharts, boxplots and scatterplots are available for displaying their corresponding data types. Special features include the display of regression lines on scatterplots and the display of 95%-level confidence intervals on boxplots.

• *Cabri-géomètre* [3] is becoming widely known as a Euclidean toolkit, which allows the construction and exploration of geometric figures and their properties from a Euclidean perspective. Any points or lines which are *created* may be dragged by the mouse, leaving intact any constructions which depend on them: it therefore offers children a way of visualising and exploring a family of *constructions* (or even the general one) by building a single example. There are now at least two variants on the idea, *Geometer's Sketchpad*, and *Geometry Inventor* (which is linked to a spreadsheet-like table-generator).

• *Tabletop* [4] is a database for children. While it incorporates advanced features (a relational version is in preparation), its central characteristic is that records are represented as (user defined) icons. These

migrate to positions on the screen defined by Venn rings which are set up by database specifications - children can visualise what the effect of a restriction on a query is in terms of where the data 'goes'. A *Junior Tabletop* version is nearing completion and this might be useful for infants.

All these programs are the result of prolonged and serious thought about mathematics and about education [5]. That does not suggest that the programs by themselves would necessarily teach children mathematics without careful intervention by teachers, the development of appropriate materials etc. Nevertheless, these programs are more than just tools, they are programs which attempt to offer children access to the key mathematical concepts within a particular domain. The computational objects within them are pretuned to particular mathematical ideas and concepts: and so we might reasonably hope that some of those ideas and concepts will resonate with children exploring that domain. *We are witnessing the rebirth of educational software.*

PROGRAMMING

In the nineteen-eighties Logo showed the way in which children could work with mathematical ideas which would otherwise be unattainable [6]. The key facet of the Logo environment has been that it offers children a way to do mathematics or in which mathematical ideas can be used as a prelude to understanding (rather than the reverse). There is enough evidence that Logo offers children the possibility of developing control over their own learning - they, rather than their teachers, ask the questions and answer them in their own way and in their own style. The teacher's role becomes subtly changed in such interactions - and just as crucial to children's mathematical development.

Logo has provided fertile ground for the construction of new microworlds - work which has been spearheaded by the ATM in the UK. Internationally, the attempt to recast mathematics from a programming perspective is gaining some ground (ten years after the publication of Abelson and diSessa's *Turtle Geometry*) with, for example, the publication of Al Cuoco's book *Investigations in Algebra*, James Clayson's *Visual Modelling with Logo*, and Philip Lewis' *Approaching Precalculus Mathematics Discretely* [7].

The development of this culture has taken longer than many expected, for a variety of social, educational and political reasons which I cannot elaborate here. On a technical level, many teachers and students, perhaps misled into thinking that Logo was just about turtle graphics, or that it made programming simple, have found that the development of Logo microworlds has not been as straightforward as they hoped. In most schools, 'doing Logo' means just that: writing simple programs (mostly to draw pictures) [8]. This is a far cry from Papert's vision of a 'richly inter connected set of microworlds', each one offering the chance to explore - through programming one or more mathematical topics. One response to this state of affairs has been to design new versions of Logo which are more than just programming languages: which offer built-in possibilities for programming *something* rather than just program ming. This has spawned the introduction of a variety of Logos such as *LogoWriter* (which combines Logo with text editing), *LogoExpress* (communications) and a new generation of Logo [9] with (simulations of) parallel processing, programmable buttons, sliders, painting tools etc.

A second response has been to develop powerful new prototypes for much more general and broadly functional computational media, of which the most interesting is *Boxer* [10] an integrated medium for programming, graphical and textual manipulation. Andrea diSessa, *Boxer's* co-creator at the University of California at Berkeley, puts it like this.

'The *Boxer* Project is aiming to test the feasibility of filling a rather grand and still hypothetical social niche, that for a computational medium. We are trying to produce a prototype of a system that extends with computational capabilities the role now played in our culture by written text. It should be a system that is used by very many people in all sorts of dif ferent ways, from the equivalent of notes in the margin, doodles and grocery lists, all the way to novels and productions that show the special genius of the author, or the concerted effort of a large and well-endowed group. In a nutshell, we wish to change the common infrastructure of knowledge presentation, manipulation and development. More modest ly, we want a general purpose system to serve the needs of students, teachers and curriculum developers, something that is so useful for such a broad range of activities that the community as a whole will judge it valuable enough to warrant the effort of learning a new and extended literacy [1 1] .

In diSessa's vision, *educational* acquires a new meaning - indeed the division between education al and commercial becomes blurred. If diSessa's vision became reality, it would be hard to define the boundaries between educational and commercial, between pedagogic and utilitarian.

Environments such as *Boxer* aim to be generally functional: they can be tuned by the learner or teacher because access is routinely gained to the tuning mechanism itself (through programming). In *Boxer*, the boundary between writing programs and, say, exploring data, is much more blurred than in traditional programming languages.

THE FUTURE

The difficulties faced by mathematics teachers are only partly technical: we know so little about how to use computers to promote deep and genuine change in schools that it would be foolhardy to attempt to choose between pretuned or tunable systems as the best means of radically reshaping mathematical learning. And in the future, we may not need to choose. Mike Eisenberg, of the Univ ersity of Colorado, is working on producing *programmable applications*, computational worlds which are finely tuned to allow exploration and problem-solution in specific domains, but which allow a completely *general* facility to (re)program. Eisenberg is trying to build on the now common place facility of any respectable application (word processor, spreadsheet, etc.) which allows the user to tinker with this or that feature, to fine-tune the programmer's pretuning. Both Eisenberg and diSessa are trying (from somewhat different directions) to get to the same place, to build software in which the learner does not have to choose between two distinct kinds of system - on the one hand, a radio which can only receive a single station, albeit providing pleasant listening and powerful reception, and on the other hand, a radio which can receive a wide range of stations, but of varying strengths and listening qualities, and with a cumbersome and inaccessible tuning knob.

Richard Noss works at the Institute of Education, University of London

This article is based on part of a paper written jointly with Celia Hoyles for the DFE conference on IT in mathematics education, February 1993.

REFERENCES

1. *Multimap* (for the Macintosh) was developed by Wally Feurzeig and Paul Horwitz at Bolt Baranek and Newman Systems and Technologies Inc. and is available for research and educational use, It does not as yet have a commercial publisher. BBN's address is 10 Moulton Street, Cambridge, Ma. 02138, USA.

2. *Elastic* (for the Macintosh) was developed at BBN on a grant funded by the National Science Foundation with Principal Investigator Andee Rubin and is available from Sunburst Publications, where it goes by the name of Statistics Workshop.

3. *Cabri-géomètre* (for the Macintosh and PC) is available in the UK from Chartwell-Bratt Ltd, Old Orchard, Bickley Road, Bromley BR1 2NE.

4. *Tabletop* (for the Macintosh) is developed by TERC and will be published as part of Jostens Learning Inc.'s ILS (integrated learning system); Educators who wish to work with a prepublication prototype can contact Chris Hancock at TERC, 2067 Mass. Ave., Cambridge, Ma. 02140, USA.

5. ...and, frustratingly for English readers, are (mostly) only available for the Macintosh. There is an old adage of educational computing which runs: 'First choose your software, then decide on your hardware'. I leave the implications of this for school purchasing policy as an exercise for the reader.

6. I used to be somewhat reluctant to make this claim, but I feel now that there is a sufficient body of evidence to support it. Some of this can be found in Hoyles C. & Noss R. (1992) (Eds.) *Learning Mathematics and Logo Cambridge*: MIT Press.

7. All published by MIT press.

8. A notable exception is the *Logo2000* pack which provides a Logo-based spreadsheet, database and graph-plotter. It was developed by Bob Ansell, Dave Pratt and Dave Wooldridge, and is published by Stanley Thornes Software as part of the *Century Maths* scheme.

9. Perversely, this is called *Microworlds*, for reasons best known to its publishers, *Logo Computer Systems Incorporated*, who, after having developed almost all of the world's Logo versions, have decided to drop the name *Logo*. Does this tell us something about the position of Logo in the North American educational culture?

10. Readers may like to refer to *Introducing Boxer* by Celia Hoyles and Richard Noss, *Micromath* 8:3, Autumn 1992.

11. A. diSessa *Social Niches* for *Future Software*, 1988 p.3, University of California, Berkeley.

CALCULATORS AND PARENTS

BARBARA HEPBURN

As part of my In-service work I was invited to support a local school at a parents evening. The school had a policy for making calculators freely available to children. However the teachers were still being questioned by parents about the need for children to use calculators instead of 'their brains'. Before the evening a large sheet was displayed in the entrance to the school and parents were invited to write down any questions they would like to ask or to make any statements about children using calculators in school. These were to give me a starting point for the evening. Predictably, the same comments appeared several times.

"How can children do mental arithmetic when they are using a calculator?"

"I did not need a calculator to do my mental arithmetic."

It was obvious that I had to convince parents that, by using a calculator, they were also doing mental arithmetic.

The evening was for parents and not for families but it was decided that we needed to show the parents how the children worked with calculators. A small group of Year six children was invited to be 'around' and work alongside the adults. The children had not worked on the activities I offered the adults. It was interesting to note that nearly all the parents brought their own calculators along with them. They all used calculators of one type or another in the course of their work and several used them to help with shopping (best value, totals etc) or with domestic accounting.

Before introducing the first activity I told the parents that, throughout the evening, I would be returning to the same questions.

"What mathematics are you working on?"

"What mental arithmetic have you been doing?"

"Would you have done that calculation using paper and pencil?"

I also suggested that they write down anything they worked out away from the calculator. They found this

difficult — one parent kept saying, "I don't need to do that on the calculator because I just know it!"

The first activity was,

> Using only the × key, what do I have to multiply 43 by to make 100?

After a few groans and some looks of panic it was heads down to some real mathematics. After a short while I stopped them working and asked, "What was the first mathematics you used?"

"I multiplied 43 by 2 in my head."

"I knew that 40 times by 2.5 is 100, so I then used the calculator to multiply 43 by 2.5."

"I divided 100 by 43 on the calculator but then I remembered you said only the times key."

(From a child) "I decided it was going to be 2 point something so I tried 2.5 because that's an easy number. It was too big so I tried 2.4. I'm going to have to change that now because I still haven't got 100..."

In response to the question, "Is there anyone who has not, so far, done any mental arithmetic?" I was greeted with nods and "I've done more mental arithmetic than calculator maths!" "I did not realise how much I was doing until you asked me!"

I then asked, "Can children use calculators without doing some mathematics in their heads?"

No-one was listening; they had returned to the problem. There was much discussion going on with both adults and children comparing methods.

I was able to observe Thomas working alongside an adult (not his parent). He was explaining how he was approaching the problem.

"43 multiplied by 2.4 is over 100, 43 by 2.3 is less than 100. I will try timesing by 2.35 because it's half way. That's 101.5."

He then does 43×2.34 (100.62)

and then 43×2.335 (100.405)

The adult asked how he decided what numbers to put in to the calculator. He paused, "I think I'm taking off point 5 every time, I'm making it point 5 smaller. I'm going to have to write the numbers down because I keep forgetting what I typed in."

Many parents and children were beginning to ask if their numbers – 100.3835...99.76 – were near enough. There was general agreement that only 100 was acceptable. I knew from past experiences of this activity that at some point in the evening some one would declare that they had the answer.

In the meantime Thomas had discovered that 2.3255×43=99.9965. I asked him how much he needed to add to this answer to make 100. "It's .35 I think... but there will be some nothings before the 35." I left Thomas working and the lady beside him was fascinated to see a child working in this way. "He knows what he is doing," she said as I moved away, "but he has difficulty explaining it to me, but that's OK isn't it?"

The evening progressed with a mixture of games using calculators and some activities looking at number patterns. There was much discussion in the plenary session about what mathematics they had been involved in during the evening. Most parents said that until they were made to stop and think about what they were doing they had not realised how much mathematical thinking they had been involved in. It was agreed that, by using a calculator, children were not taking the easy option. In fact they were able to become involved in mathematical thinking which in the past

had been cluttered by having to do paper and pencil calculations at a very low level. Here was a group of children confidently handling numbers to five decimal places. The calculator was able to remove the drudgery from the mathematics. (It was during the plenary session that an adult claimed to have an answer to the first problem.)

Some of the parents expressed surprise at the sort of things the children were able to do. Most of them thought that the children just used the calculator to check their answers or even to give them the answers to the usual 'sums'. Some of the children were quick to tell the parents that they did not 'cheat' by using a calculator. One girl said, "We use a calculator sensibly".

In her article *Let's stop wasting time* (Micromath, Spring '92), Hilary Shuard encourages us to give the parents who oppose the use of calculators 'the opportunity to observe, in the primary classrooms, how the children can use calculators to enrich, widen and develop their mathematics.' I do not think that observation is enough. My experience has shown that parents need to work on some activities for themselves alongside children. They will then see, as the parents in this Cumbrian school did, that using calculators can enrich and widen their children's mathematics.

About a week after the parents' evening I received the following from one of the children ...

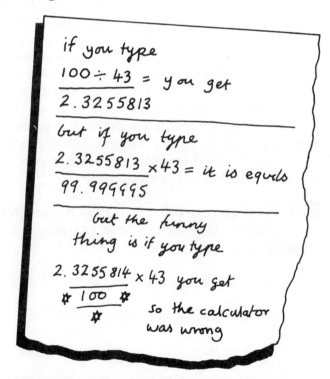

if you type
$$100 \div 43 = \text{you get}$$
2.3255813

but if you type
$$2.3255813 \times 43 = \text{it is equals}$$
99.999995

but the funny thing is if you type
$$2.3255814 \times 43 \text{ you get}$$
✿ 100 ✿
✿
so the calculator was wrong

Barbara Hepburn works at Charlotte Mason College, Ambleside.

DIMENSIONS OF ASSESSMENT

Many people seem to think that written tests are necessary to pupils' progress. **Tandy Clausen** has other ideas.

The holy grail of mathematics education is a test. A nice, short, easy-to-administer test, which we can use to establish the exact level of attainment and the potential of each pupil in the class.

It does not exist, of course. Tests are notoriously inaccurate in their results – and the more 'objective' they purport to be, the more those results will reflect factors other than the pupil's mathematical understanding. Formal written tests provide a measure of many things – of the pupil's ability to sit quietly in an exam room, of their ability to read and write, and their ability to work out what particular technique or item of knowledge the examiner is after in each question. But a pupil who cannot, for whatever reason, handle the formal test situation, will fail the test – and so be regarded as a failure in mathematics.

But teachers and assessors will go on using formal tests, and the search for a 'good' one will continue. The search will continue because we need one so badly. How else can we assess our pupils effectively – especially new arrivals to the classroom, about whom, in practice, little may be known, and who need to be fitted in to the class structure as quickly and as painlessly as possible?

Working in a special school for emotionally and behaviourally disturbed children I face this problem regularly. The reality is that many pupils arrive in my classroom with little warning and inadequate records, often in a very negative frame of mind. They know that they failed at their previous school or unit, and they may have been out of school for weeks, months, or occasionally years. In some cases they have recently been through the courts. Asking these pupils to do a formal written test would be pointless, and perhaps even harmful. Their response might well serve to demonstrate that they are severely disturbed – but we know that already. We want to find out what they can do, not what they cannot.

So how do we go about assessing the mathematical ability of such pupils? I asked myself and some colleagues in special education this question recently, and we came up with some interesting answers.

We had noticed that many of our pupils whose performance with literacy and with written number work was poor were surprisingly good at practical spatial activities. Cockcroft, over ten years ago, observed that what was then called 'remedial' education tended to focus primarily on the development of the pupils' literacy skills, which led to an 'excessively narrow' mathematics curriculum, in which 'there was a tendency to restrict the courses provided for the less able to routine calculation divorced from context' (Cockcroft et al: *Mathematics counts*, p143, 164.) The introduction of GCSE, and of the National Curriculum with its clear emphasis on other areas of mathematics, is doing much to encourage a wider curriculum at all levels, but it is still the case that the dependence of mathematical achievement upon literacy skills creates problems for many of our pupils. On the other hand, a non-written approach can be much more successful.

One teacher I talked to works in an assessment unit for pupils with a range of problems, from disaffection and school phobia to severe disturbance. She gives newcomers the SMILE *Red cube* to play with. This is a wooden 3 by 3 by 3 cube, made up of twenty-seven unit cubes. The outside of the 3 by 3 by 3 cube is painted red – so some faces of some of the unit cubes are painted, while others are plain.

My colleague usually starts by asking the pupil to put the twenty-seven unit cubes together into a block, without focusing on the painted and unpainted faces. Some pupils work by trial and error, trying various numbers of cubes in the bottom layer and then adding more layers. These pupils realise that each layer must be made up of a rectangle of unit cubes, and that all the layers must have the same number of cubes in them, but they are unable to use this knowledge to work out how many layers there are or how many cubes in each. This indicates that the pupil is working at level 2 or thereabouts, in both *Number* and *Shape and space*.

Other pupils, who have a good grasp of the concepts of multiplication and division – even if they do not know the tables off by heart – will quickly realize that the twenty-seven unit cubes can only be divided into three layers, and that the nine cubes in each layer must form a 3 by 3 square. These pupils are generally working at about level 4.

Pupils who go on to make the red cube may be working at a higher level. For example, the pupil may recognize that the eight corner unit cubes must have three sides painted, while the edge cubes have two and the 'middle of the face' cubes have one. A pupil who works this out without help may be ready to tackle other mathematics at level 6: if a little help is needed then they are probably at level 5.

It is very easy to pick holes in this strategy for assessing pupils. It does not really work by rule of thumb, as I have perhaps described it here: assessment through the careful observation of a practical task is a very personal interactive process, in which the teacher focuses closely on the pupil, and the pupil, relaxed and ready to co-operate, explores the ideas offered. But it works: this approach to assessment provides far more useful information in a short space of time about the pupil's mathematical ability than we could ever hope to obtain from a formal written test.

We each have our own favourite assessment tool: mine is *Shape fit*, one of the range of foam rubber DIME materials which have motivated and enabled so many of my pupils in the mathematics classroom. Each pupil is provided with a set of ten tiles and a series of puzzles printed on card. The tiles must be fitted on to the outlines shown in the puzzles. On the early cards (numbers 1 to 4) the whole outline of each tile is given, so the puzzle is a simple matching task. Even this, however, requires the pupil to rotate and reflect the tiles to fit them to their outlines. Pupils working at about level 2 can usually achieve this, but the readiness with which they select and place the tiles provides an indication of their achievement beyond this level.

Cards 5 to 10 provide more difficult tasks: the tiles are arranged in pairs, and the joining lines are not drawn in – only the outlines of the complete shapes are given. These present a fair challenge to pupils working at level 3. Then the puzzles get progressively harder with more tiles being grouped together with no joining lines: a pupil who, without too much difficulty, can tackle the last four or five of the twenty cards provided is likely to be able to work at level 4 or 5 in mathematics.

Again, the 'rule of thumb' approach is patently absurd. And yet teachers of pupils with special needs do develop strategies like this in order to assess their pupils quickly, accurately enough, and in an enjoyable, unthreatening way. The activities described here are certainly not enough in themselves – but they provide a good start!

An assessment activity like *Shape fit* has the added advantage that it leads quite naturally to a more open-ended situation in which pupils design their own puzzles and test them out on each other. This idea started in a quiet way at our school with Nicholas, a boy in Year 7 who was with us for only a short period before making a successful transfer to a mainstream school. Nicholas had worked with great enthusiasm through the cards provided in the pack, and he wanted to do some more. There weren't any, so he made one. The other pupils in Nicholas' class were not particularly impressed, but a few weeks later Charlie, a very bright but severely disturbed Year 8 boy, reached the end of the cards provided in the pack. I offered him Nicholas', without telling him where it came from: he found it a challenge, but was able to solve the puzzle eventually. When he realized that this had been produced by another pupil, Charlie decided to try his hand himself at puzzle design. Pat, the special support assistant who always worked with the class joined in, and then Clyde, who could not read but could draw very well, was inspired to try. Cards number 21 to 24 in our version of the *Shape fit* pack have the names of Nicholas, Charlie, Pat and Clyde at the top – and, with time, another thirty-odd cards have joined the set. Even Joe, who could not control a pencil very well, was able to design a puzzle, which his helper drew around under Joe's direction.

As the number of new cards increased, interest grew in their relative difficulty. Cards 1 to 20 – the ones which came with the pack – were more or less in order of difficulty, but this was clearly not the case with the school-produced cards. Two boys, Christopher and Clyde, set out to rank the new cards, giving each puzzle a mark out of ten, depending how hard it was to solve. They worked independently, and then compared their judgements: they were pleased to find that to a great extent they agreed. This led to some discussion about what makes a 'difficult' or an 'easy card': obviously, the more tiles there are grouped together the more difficult the card may be, but if all the tiles are used to make two different designs one may still be easier than the other. The boys concluded that some designs allow for more choices during their solution, and this makes them harder.

The importance of motivation to mathematical achievement has been often remarked. It is even more significant in the context of emotionally damaged children like many of ours, who have learnt that they must fail and will not take the risks involved in trying to succeed. Relaxed, informal, practical tasks such as the ones described here can help us, as teachers, to give back to our pupils the confidence to enjoy mathematics, and to take it for their own.

Tandy Clausen taught at the Jonathan Miller School, Slough, when she wrote this article. Now she works at Kings College, London.

TESTING VERSUS ASSESSMENT

MIKE OLLERTON

In recent years mathematics education has seen some positive moves away from tight didactic teaching methods, where children were taught concepts without an understanding of what they were doing, or without a context on which to base their learning, to a problem solving approach. Because this latter type of approach parallels the way learning takes place in the 'real world' it therefore should underpin the way that children learn mathematics effectively.

The changes that have happened in the classroom must also have consequences for the way that a child's performance is measured. The kind of changes to which I refer support the aims of mathematics education which were first recognised at a national level through Cockcroft in 1981, supported and encouraged in 1985 by HMI *(Mathematics from 5 - 16)*, developed in schools through the introduction of coursework in some GCSE syllabi since 1986, and institutionalised by the National Curriculum NSG in 1990.

Teachers of mathematics are currently striving to incorporate these ideals into their teaching methodology. Unfortunately our confidence as teachers to enhance and enrich our students learning of mathematics is often met by ignorance, prejudice, rhetoric and contradiction.

Because the National Curriculum document for Mathematics has been written at 10 levels over 14 Attainment Targets, there is a fundamental danger that teaching methods will focus on narrow pieces of content:

> *Much of the mathematical experience of most pupils is extremely fragmented ... Indeed because of the commonly held view that 'many pupils cannot concentrate for any length of time' many text books are planned to provide this rapidly changing experience. However provided that topics which interest them are selected it is possible to encourage most if not all pupils to pursue a study in some depth ... An in-depth study is of potential value for all pupils, not only mathematically but also in terms of development of personal qualities such as commit ment and persistence.*
> Mathematics from 5 -16, HMI

Consequently the assessment of the students' responses may be narrow and fragmented and something that happens only at the end of a piece of work, usually through an end of topic test. I believe that a more holistic approach is necessary and more valuable. I want teachers to develop problem solving attitudes in their students so that assessment of the content described in Attainment Targets 2 - 8 and 10 - 14 is made by applying the level related criteria of the process based Attain ment Targets 1 and 9. This can be achieved by focusing on assessment using broad criteria across a range of content. The type of project work undertaken in secondary schools through GCSE coursework assessment needs highlighting as examples of good practice.

The mathematics department at Orleton Park School has begun to develop similar techniques enabling children in the 11 to 14 age range to produce 'coursework' type responses, and then make appropriate comments back to the student as an ongoing part of the assessment.

The easiest objectives to assess are facts and skills but any assessment is inadequate if these are all that are assessed . Due attention needs to be given to the assessment of conceptual structures and general strategies because these objectives are more indicative of pupils' mathematical abilities.
> Mathematics from 5 -16, HMI

It is relatively easy to access facts and skills out of con text, but this has little value. With regard to a particular skill it is not enough to ask whether the pupils can perform the skill; we must also ask whether they can use it in a variety of contexts, whether they understand it ...
> Mathematics from 5 -16, HMI

It is neither possible nor desirable to teach in one fashion and then test the students in a different way. I believe that narrow testing of the kind prevalent in GCSE end of course examinations can only assess narrow mathematical skills; it will not assess any of the richer areas of mathematics which are fundamental to good practice, such as:

- *appreciation of relationships within mathematics*
- *imagination, initiative and flexibility of mind in mathematics*
- *working in a systematic way*
- *discussion*
- *problem solving*
- *appropriate practical work*
- *investigational work*

Mathematics from 5 -16, HMI

I would add to this list the following set of skills that also cannot be tested in the traditional manner; these are the ability to:

- *communicate an extended piece of mathematics*
- *implement ideas behind an extended piece of mathematics*
- *process and interpret information requiring skills such as classifying, generalising, hypothesising and proving.*

The result from a narrow content without context test does not reveal information about whether a student could achieve the same result a day, a week or a month later. It would not tell the teacher that the student could apply the tested knowledge in a variety of different contexts outside the sterile environment of the examination room. All that such a test will inform anybody is that at a particular moment in time the student was, or was not able to respond to particular questions written by someone else. There are of course dangerous consequences, such as what the teacher then does with this information. Do they re-teach the skills from the test that the student has already failed to get to grips with? Does the teaching method revert back to a narrow skills based technique so that the student can answer a similar test question (but with different numbers) on the next occasion?

At present (July 1985) undue importance is often attached to termly or yearly tests. If these produce any major surprises for the teacher it reveals that the assessment aspect of the teaching approach is inadequate 'because effective teaching can only take place with continuing assessment of pupils' responses.'

Mathematics from 5 -16, HMI

A test cannot assess the way an individual carries out mathematical activity that draws upon the student's autonomy, or mathematical attitude or their ability to evaluate the work they do; as putting the student in a

test situation defeats autonomous action and denies opportunities to display either a spirit of enquiry or reflect upon a piece of mathematics.

'These (Objectives 23 and 24) need to be assessed but mainly in an informal way... If these personal qualities are missing then there is something fundamentally wrong with the mathematics curriculum whatever the levels of achievement attained by the pupils.

Mathematics from 5 -16, HMI

A test in itself does not raise standards. At its worse a test will undermine confidence in all but the most able and will therefore reinforce failure amongst the majority. A test can however prevent and therefore undermine a student's opportunities to demonstrate their ability to function as a mathematician.

If students are actively encouraged to perceive that the important aspects of mathematics are narrow skills that are immediately testable, it will inhibit their confidence to recognise the value of and therefore participate more actively and willingly in the broader aspects of mathematics, as listed above. This will in turn have a retrograde effect upon students' mathematical aspirations.

At various points in this document comments are made about the undesirability of overemphasising the practising and testing of skills out of context; the ability to carry out operations is important but there is a danger that skills come to be seen as ends in themselves. y mathematics is only about 'computational skills out of context' it cannot be justified as a subject in the curriculum.

Mathematics from 5 - 16, HMI

For many years, mathematics has largely been taught in 'bits', usually in a linear fashion. The difficulty that arose from this kind of learning experience is that the student did not often have the facility or the understandings that are required to join the bits together. Testing will foster the re introduction of 'bit-teaching' and deny developing understanding as a whole.

When presented with a mathematical task pupils should be encouraged to find their own method of carrying it out even though there may be a standard more streamlined method which they might ultimately learn... Many textbooks present problems' which use words and give the exact amount of information required for solution. These questions require pupils to make direct use of a skill which has just been learnt, but they provide relatively little real challenge. Opportunities need to be given to pupils to use their expertise to find

*their own way through problems and investigations
... The aim should be to show mathematics as a
process, as a creative activity in which pupils can
be fully involved, not as an imposed body of
knowledge immune to any change or development.*

Mathematics from 5 -16, HMI

It is essential that as the students progress through the school they take evermore responsibility for their learning in order to achieve independence. To achieve this, tasks should be both open-ended problems or investigations and content-based courseworks, thus enabling differentiation by out come. Investigatory ways of working are not just restricted to open-ended problems, these skills are constantly encouraged throughout the more directly content based components of the mathematics syllabus, such as trigonometry and Pythagoras.

*Pupils should have opportunities to explore and
appreciate the structure of mathematics itself ...
Mathematics is not only taught because it is useful.
It should also be a source of delight and wonder,
offering pupils intellectual excitement and an
appreciation of its essential creativity.*

Mathematics - The NCC NSG

*Learning in mathematics does not necessarily take
place in completely predetermined sequences.
Mathematics is a structure composed of a whole
network of concepts and relationships, and, when
being used~ mathematics becomes a living process
of creative activity.*

Mathematics - The NCC NSG

For the past 5 years I have worked in a school that has been developing assessment of coursework through a GCSE syllabus written by members of the Association of Teachers of Mathematics, and certificated through the Southern Examining Group. There are six other schools operating the same syllabus. The style of teaching and learning that this syllabus encourages has produced some interesting developments, such as paving the way for mixed ability teaching groups through to age 16. There has also been a noticeable improvement in the performance of girls in mathematics. Students are encouraged to behave as mathematicians, working on problems which offer them opportunities to explore and research ideas that help them understand and work with traditional mathematical concepts. Because a high expectation is placed upon the students their responses to the ideas that they work on is of a high standard. This work is then assessed using meaningful criteria. Accordingly the students motivation is higher, their learning is enhanced, and I firmly believe that they raise their own standards. The type of teaching and learning experiences that have proved to be successful for helping to record assessments for the GCSE are now

being developed in the lower school in preparation for a KS3 assessment model.

This experience leads me to the conclusion that useful assessment must be of a continuous nature and not by simple testing.

*Orleton Park School, Telford
University of Keele (Joint Appointment)*

Changing image

Mathematics subject assignment: *Share with us some mathematics on which you have engaged this term. Write an account of your findings and, at the same time of the experience of carrying out the task; insights, frustrations, conversations, crises and satisfactions. How would you/have you introduce(d) the ideas to children? How does working on the mathematics for yourself inform your practice?*

Throughout my entire mathematical education this has been the first time that anyone has ever suggested to or encouraged me to look at some mathematics that interests me, and given me the responsibility of choosing it. In just about every other subject I have been given choices in what I want to study or do eg humanities, art, CDT, science and health education projects. So, before I even started my assignment, the actual thought of it raised an important issue for me.

– Why, throughout my entire education, in which mathematics has always been my favourite subject, have I never looked at or chosen my own topic of study?

– Is it right that any mathematics has been restricted or limited to the choices my teachers or tutors have made?

– Is this difference in approach between subjects a contributory factor in many people's hate or misunderstanding of mathematics?

– Do people feel that, unless a teacher is present to teach or explain, then they can't do mathematics ie maths cannot be 'done' outside the classroom?

Sara Stratton,

On main school practice, The Ridings High School, Winterbourne

Students know when I am bored with a topic, or nervous, or uncertain. The unscrupulous can exploit these perceptions by being very effective at disrupting the proceedings. Somehow, they can find my weak spots quickly. If I am promoting a questioning, explorative approach, promoting mathematics as a disciplined form of enquiry and as a formalised expression of awarenesses that everyone who speaks and walks must necessarily have, then it seems to me that I must myself be in such a state when teaching. It is difficult to be genuinely surprised by an answer to a routine question that students have worked on year after year, but it is possible to be genuinely surprised and excited about the processes which students invoke, about what they have to say about their own thinking. I find myself getting excited in almost every session I run, and when I do not it is a disaster. But, my excitement is about being aware in the moment of recognising mathematical thinking, the use of imagery, the evocation of basic powers to make sense of things – and of possibilities to exploit what I notice and draw attention to mathematically significant moments.

John Mason

Centre for Mathematics Education,
The Open University